AA003727

12th International Beam Instrumentation Conference 2023

Journal of Physics: Conference Series Volume 2681

Saskatoon, Canada
11-14 September 2023

ISBN: 978-1-7138-8961-8
ISSN: 1742-6588

Printed from e-media with permission by:

Curran Associates, Inc.
57 Morehouse Lane
Red Hook, NY 12571

CURRAN ASSOCIATES INC. proceedings.com

Some format issues inherent in the e-media version may also appear in this print version.

Institute of Physics
Dirac House, Temple Back
Bristol BS1 6BE UK

Phone: 44 1 17 929 7481
Fax: 44 1 17 920 0979

techtracking@iop.org

Additional copies of this publication are available from:

Curran Associates, Inc.
57 Morehouse Lane
Red Hook, NY 12571 USA
Phone: 845-758-0400
Fax: 845-758-2633
Email: curran@proceedings.com
Web: www.proceedings.com

TABLE OF CONTENTS

Preface

The 12th International Beam Instrumentation Conference, IBIC'23, was hosted by the Canadian Light Source (CLS) in Saskatoon, Saskatchewan, Canada from Sunday, September 10th to September 14th, 2023. IBIC is dedicated to exploring the physics and engineering challenges of beam diagnostic and measurement techniques for particle accelerators. IBIC'23 was attended by 182 delegates from 18 countries. IBIC'23 included 18 invited talks, 18 contributed talks, 2 tutorials, 114 posters, and a 3-day vendor exhibition with 18 industrial exhibitors in attendance. Twenty-five young scientists from all over the globe attended the conference. Nine of these students received grants thanks to sponsorship from the City of Saskatoon, Tourism SK, Discover Saskatoon and Innovation SK and contribution from the industrial exhibitors. The organizers of IBIC'23 are grateful to all sponsors for their valued support.

The conference venue was TCU Place, a 104,000-square-foot convention and arts centre situated in downtown Saskatoon. Welcome remarks were delivered by CLS CEO, Bill Matiko, Saskatoon Mayor, Charlie Clark and University of Saskatchewan President Peter Stoicheff and Tonia Batten (CLS), IBIC'23 Conference Chair, opened the conference.

The scientific program was developed by a 23-member Scientific Program Committee (SPC). Oral and poster sessions were grouped according to the following nine IBIC classifications.

Beam Charge and Current Monitors

- Diagnostic systems measuring the average beam current, instantaneous total intensity or individual bunch intensity.

Beam Loss Monitors and Machine Protection

- Diagnostic systems measuring the average beam loss, instantaneous beam loss or individual bunch loss.
- Machine protection architectures based on inputs from beam instrumentation systems, as well as machine protection apparatus such as scrapers, collimators, radiation shields, etc.

Beam Position Monitors

- Diagnostic systems used for the measurement of beam position: orbit measurement techniques, trajectory measurement techniques, and bunch-by-bunch position measurement systems.
- This track includes the use of diagnostics associated with secondary beams, e.g. photon beamlines, to measure the position of the charged particle beam.

Data Acquisition and Processing Platforms

- Data acquisition systems or architectures of relevance for beam diagnostic systems. This includes acquisition platforms, data processing techniques, electronic component validation or characterization.
- Computing platforms, control and acquisition software and operating systems of relevance for beam diagnostics.

Feedback Systems and Beam Stability

- Systems used to stabilize or control any beam or accelerator parameter either in a closed or open loop. This includes control of orbit, trajectory, longitudinal stability, bunch-by-bunch transverse stability, emittance, tune, chromaticity, etc.
- This can include descriptions of the detectors, actuators, any specialized acquisition electronics used to acquire the signal, as well as related data processing techniques. The feedback or feedforward aspects should, however, form a major part of the submission, otherwise, the diagnostic description would be better suited to the track dealing with the primary measurement under discussion.

Longitudinal Diagnostics and Synchronization

- Diagnostic systems used for the measurement of longitudinal beam parameters such as bunch length, bunch profile, arrival time, energy spread, or longitudinal emittance.
- Systems used for beam synchronization and timing distribution.

Machine Parameter Measurements

- Systems used to measure machine parameters such as betatron tune, chromaticity, higher-order multipoles, space charge, impedance, beam-beam, etc.
- Techniques and algorithms used for such measurements.
- Other diagnostic devices or techniques which do not clearly match any of the previous tracks.

Overview and Commissioning

- General description of beam instrumentation devices and systems at new facilities or machine upgrades.
- Commissioning results and lessons learned.
- Workshop reports related to beam diagnostics.

Transverse Profile and Emittance Monitors

- Diagnostic systems used for the measurement of transverse beam size, transverse beam profile, transverse emittance, and beam halos.
- This track includes both interceptive and non-interceptive diagnostics.

The distribution of talks by classification was 5% Beam Charge and Current Monitors, 8% Beam Loss Monitors and Machine Protection, 13% Beam Position Monitors, 16% Data Acquisition and Processing Platforms, 8% Feedback Systems and Beam Stability, 18% Longitudinal Diagnostics and Synchronization, 8% Machine Parameter Measurements, 16% Overview and Commissioning and 8% Transverse Profile and Emittance Monitors.

The regional distribution for participation was 37% from the Americas, 18% from Asia and 45% from Europe. The regional distribution of talks was 32% from the Americas, 26% from Asia and 42% from Europe. The gender ratio for oral presentations was 82% male and 18% female. The oral presentations were 79% onsite talks and 21% remote talks.

Journal of Physics: Conference Series **2681** (2024) 011001 doi:10.1088/1742-6596/2681/1/011001

The scientific program was supplemented by a variety of special events.

The welcome reception was hosted at Remai Modern. The Remai Modern is an internationally acclaimed are gallery located on the bank of the South Saskatchewan river which provides both Canadian and Saskatchewan perspective on global modern and contemporary art movements through a mix of regional, national and international programs.

The conference banquet was hosted at Champêtre County. This is a wild west themed special event venue offers catering and themed activities that will leave you with an inner peace, having experienced the simpler things in life, while learning about Saskatchewan and the prairies in a unique way.

There was also a tour of the Canadian Light Source (CLS), which is Canada's national synchrotron light source facility, located on the grounds of the University of Saskatchewan. The CLS is a leading centre for research excellence with a focus on health, agriculture, environment, and advanced materials. Since the start of operations, in 2005, the CLS has enabled over 5,000 scientists from 56 Canadian academic institutions and 45 countries, to publish over 6,000 scientific papers, highlighting discoveries in a wide variety of fields, in over 1,000 international scientific collaborations.

Industrial Exhibition

The industrial exhibition took place during the first three days of the conference. Industrial exhibitors (10 companies) occupied 9 booths, at which they presented their high technology products and services to the delegates in an excellent atmosphere conducive to discussions wherein exhibits and posters were intermingled. The Industrial Exhibitors were also given an opportunity to introduce themselves to the entire audience in the conference auditorium. It must be acknowledged that the conference would not be possible in its present format without the generous support of the IBIC industrial exhibitors and sponsors.

Peer Reviewed Proceedings

A new feature for IBIC, introduced on a trial basis at IBIC'23, is a light peer review of a limited number of papers for publication in a volume of an Institute of Physics conference proceedings. The following white paper defining the review criteria was developed. This closely mirrors the process developed by the International Particle Accelerator Conference (IPAC) series for light peer review. Thank you to Todd Satogata, Nicolas Delerue, Nawin Juntong, Alex Bogacz and Christine Petit-Jean-Genaz for the valuable insight and direction provided.

Papers were reviewed by the Scientific Program Committee, while the Scientific Publication Board (SPB) performed the function of the editorial board. Candidate papers were submitted 7 weeks in advance of the normal deadline to allow a cycle of review, revision and final review. 12 papers were registered for light peer review, 11 were submitted, many were revised, 9 were approved. The SPB managed the review process within the Joint Accelerator Conference on the Web (JACoW) Scientific Program Management System (SPMS) database.

IBIC2023, Peer Review General Policies

Papers will be rejected as refereed papers if a submission deadline is missed, or if the review process is not completed, or if maximum capacity has been reached. The paper will then be published as an unrefereed paper at the JACoW website (www.jacow.org). Only one review is allowed in general.

- Papers will be rejected if a submission or correction deadline is missed.
- Papers are rejected if the review process is not completed before September 6th, 2023, the deadline for paper submission in SPMS to allow for the JACoW editorial team to start reviews.
- Each paper shall have two (2) reviewers.
- Single-blind review (the reviewers are unknown to the authors).
- A reviewer shall not know the identity of the other reviewer assigned to the paper.
- A reviewer shall not contact an author directly concerning their paper.
- Reviewers shall not make any personal remarks, or comments that may betray their identity, when entering the instructions for requested correction/revision of the paper.
- Reviewers must not referee papers of which they are authors or co-authors.
- If there are insufficient resources, and if the paper emanates from a different department, then a reviewer may referee papers from their home institution. This will be avoided where possible.
- Only a single cycle of correction(s) by the author(s) is permitted; such cycle shall include the opportunity to respond to correction requests from both reviewers.
- For each referee of a paper, there are only three possible outcomes from the first cycle of review:
    - Accept without correction
    - Accept with correction
    - Reject
- For each referee of a paper, there are only two possible outcomes from the second cycle of review:
    - Accept
    - Reject
- Decisions of the SPB shall be final.

IBIC 2023 Peer Review Acceptance Criteria

Statement of Peer Review:

All papers published in this volume of Journal of Physics: Conference Series have been peer reviewed through processes administered by the Scientific Publication Board. Reviews were conducted by expert referees to the professional and scientific standards expected of a proceedings journal published by IOP Publishing.

The IBIC 2023 peer review criteria are:

1) The published work must **not contain clear errors or important factual mistakes**.

2) The paper **must include own work**, performed by the authors and not published elsewhere.

3) Some aspect or part of the work must be **original** or demonstrate clear progress over other reports of the work.

4) The presentation of the results **must be understandable**.

5) The paper must be in **good English**.

6) Work and related **results by others must be referenced and properly acknowledged**.

7) The paper must include **references to literature** that are appropriate.

8) Papers are considered **not-correctable** and therefore rejected in following cases:

a) The whole Ansatz is wrong. (i.e. lacks scientific merit)

b) The work is from somebody else or claiming authorship

c) Requested changes are not implemented in time.

IBIC 2023 Questions posed to referee in SPMS review module

Reviewer must acknowledge they have read the referee guidance statements.

Reviewer answers Yes/No to the following queries:

1) Does this paper have sufficient scientific merit to be published in the IBIC'23 IOP light peer review proceedings?

2) Is the work free of obvious scientific or technical errors and important factual mistakes?

3) Does the paper include own work, performed by the authors and not published elsewhere?

4) Is some aspect or part of the work original or demonstrate clear progress?

5) Is the presentation of methods, results and conclusion understandable?

6) Is the paper written in good English?

7) Does the paper acknowledge the work of others in the same field?

8) Does the paper include appropriate references to the literature?

Outcome of each question is limited to Yes/No (entered by the referee).

Outcome of the referee dialogue

- If answers to 1) or 2) or 3) or 4) are negative (No), then paper is uncorrectable (automatic rejection).
- If answers to all questions 1)-8) are positive (Yes), then paper is automatically accepted.
- If answers to 1), 2), 3), 4) are positive, but one or more of 5), 6), 7), 8) are negative then two cases exist:
    - Easy/straight-forward to correct: then revise.
    - Difficult/complex to correct: then reject.

Reviewer then enters final appraisal SPMS Question:

Is the paper acceptable for publication in its present form?

Answer is limited to three possibilities (entered by the reviewer):

1. Yes (accept).
2. No, but easily corrected (request author to make revisions). In this case the reviewer enters instructions to the author for correction/revision.
3. No, and uncorrectable (reject).

Journal of Physics: Conference Series **2681** (2024) 011001 doi:10.1088/1742-6596/2681/1/011001

Committees

The success of IBIC'23 was a result of the strong international and local collaboration teams which include, but are not limited to the following:

Scientific Program Committee
Tonia Batten, CLS, Canada – Chair
Willem Blokland, SNS/ORNL, USA
Lorraine Bobb, Diamond, UK
Peter Forck, GSI, Germany
Dave Gassner, BNL, USA
Nicolas Hubert, Soleil, France
Kevin Jordan, JLAB, USA
Changbum Kim, PAL, Korea
Prapong Klysubun, SLRI, Thailand
Patrick Krejcik, SLAC, USA
Thibaut Lefevre, CERN, Switzerland
Yongbin Leng, NSRL/USTC, China
Shengli Liu, TRIUMF, Canada
Hirokazu Maesaka, SPring-8, Japan
Sergio Marques, LNLS, Brazil
Kenichiro Sato, J-PARC, Japan
Victor Scarpine, Fermilab, USA
Volker Schlott, PSI, Switzerland
Adriana Wawrzyniak, Solaris, Poland
Carsten Welsch, UOL, UK
Kay Wittenburg, DESY, Germany
Junxia Wu, IMP, China
Junhui Yue, IHEP, China

Local Organizing Committee
Tonia Batten, CLS, Canada
Grant Bilbrough, CLS, Canada
Mike Bree, CLS, Canada
Shawn Carriere, CLS, Canada
Darren Hunter, CLS, Canada
Ginette LaVoie - Chair, CLS, Canada
Adam Leontowich, CLS, Canada
Sandra Ribeiro, CLS, Canada
Michael Smith, CLS, Canada
Robby Tanner, CLS, Canada
Glen Wright, CLS, Canada

JACoW Editors
Michael Bree, CLS, Canada - Editor in Chief
Darren Hunter, CLS, Canada – Scientific Secretariat
Grzegorz Kowalski, GSI, Germany
Christine Petit-Jean-Genaz, CERN (Ret.), Switzerland
Volker Schaa, GSI (Ret.), Germany
Ana Štajminger, ELI, Czech Republic

Scientific Program Board
Tonia Batten, CLS, Canada
Michael Bree, CLS, Canada
Darren Hunter, CLS, Canada

IBIC'23 Peer Review Proceedings "Preface" prepared by:
Tonia Batten, Canadian Light Source, IBIC'23 Conference Chair

Journal of Physics: Conference Series **2681** (2024) 011002 doi:10.1088/1742-6596/2681/1/011002

Peer Review Statement

All papers published in this volume have been reviewed through processes administered by the Editors. Reviews were conducted by expert referees to the professional and scientific standards expected of a proceedings journal published by IOP Publishing.

- **Type of peer review:** Single Anonymous
- **Conference submission management system:** Morressier
- **Number of submissions received:** 9
- **Number of submissions sent for review:** 9
- **Number of submissions accepted:** 9
- **Acceptance Rate (Submissions Accepted / Submissions Received × 100):** 100
- **Average number of reviews per paper:** 2.3333333333333335
- **Total number of reviewers involved:** 3
- **Contact person for queries:**
  Name: Tonia Batten
  Email: tonia.batten@lightsource.ca
  Affiliation: Canadian Light Source Inc

Low intensity beam current measurement of the Associated Proton Beam Line at CSNS

R Y Qiu, W L Huang*, F Li, L Zeng, M A Rehman, M Y Liu, Q R Liu, R J Yang, T Yang, Z H Xu, Z X Tan

Institute of High Energy Physics, Chinese Academy of Sciences, Beijing, China

Spallation Neutron Source Science Center, Dongguan, China

huangwei@ihep.ac.cn

Abstract. The Associated Proton beam Experiment Platform (APEP) beamline is the first proton irradiation facility to use naturally-stripped protons which come from H- beams interacting with the residual gas in the linac beampipe at CSNS. The stripped beam current, which is in the order of 0.1% of the original H- beam and approximately 10 μA, should be measured precisely to provide the proton number for irradiation experiments. Therefore, a low-intensity beam current measurement system was developed with considerations to eliminate the external interferences. An anti-interference design is adopted in this system with an elaboration of probes, cables and electronic low-noise technology to minimize the impact of environmental noise and interferences. This improves the signal-to-noise ratio and enables a more precise measurement of the microampere level pulsed beam current. The system was installed and tested during the summer maintenance in 2021 and 2022. It shows a good agreement with the measurement of the Faraday cup.

1. Introduction

The linear accelerator of China Spallation Neutron Source (CSNS) consists of a negative hydrogen ion source, a 3 MeV radio frequency quadrupole accelerator (RFQ), and a 4-tank drift tube linear accelerator (DTL) [1]. The negative hydrogen ion beam is accelerated to 80 MeV and then injected into the Rapid Cycling Synchrotron (RCS) by charge-exchange stripping. The proton beam is then accelerated to 1.6 GeV and finally extracted to bombard the tungsten target for neutron production with a repetition rate of 25 Hz [2]. In the linac of CSNS, a portion of H- beam interacts with the residual gas. As a result, in the beampipe there are three types of particles with different charge states: H- (~15 mA, 100~500 μs), hydrogen atom (H0), and proton (H+). The associated proton beam current is about 0.1% of the original H- beam. Therefore, the APEP beam line is established from the bending magnet at the end of linac. The number of protons in the beam is a crucial parameter for irradiation experiments.

Figure 1 shows the layout diagram of the APEP area. There are three devices used for measuring beam intensty. The first one is the current transformer measurement system located at the entrance of APEP. In order to precisely measure this microampere-level beam intensity, a low-intensity beam current measurement system was developed in 2021. It achieved stable measurements by addressing external interferences in September 2022.The second device, located behind collimator 3, is the beam intensity measurement system based on the secondary electron emission method. The third device is the Faraday cup beam intensity measurement device located in front of the DUMP. The two devices behind

Journal of Physics: Conference Series **2681** (2024) 012001 doi:10.1088/1742-6596/2681/1/012001

are currently undergoing debugging. This article primarily focuses on the design of the microampere-level beam intensity measurement system for the beam current transformer.

Table 1. Parameters of the Associated Proton Beam

Parameter	Value
Proton Beam Energy	80 MeV
Repetition Frequency	25 Hz
Number of Particles	~5.0·1010 ppp@100 kW
Macro Bunch Length	100 μs ~ 500 μs
Proton Beam Intensity	20 μA@400 μs,100 kW

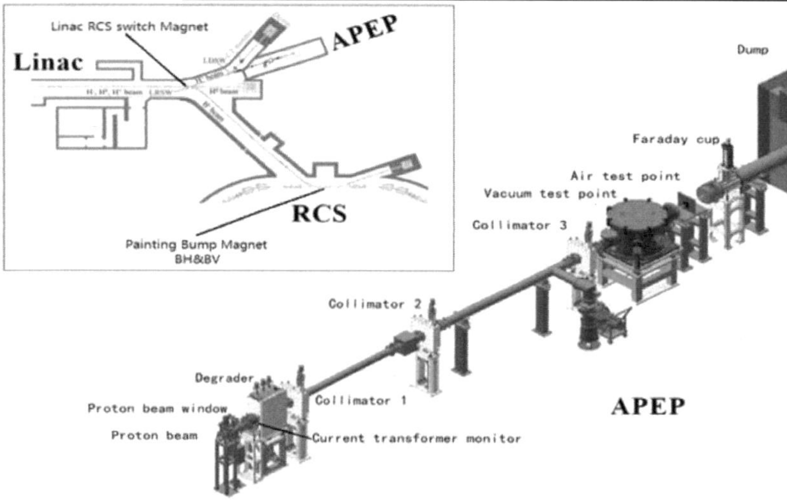

Figure 1. Layout of the APEP beam line [3].

2. System design

2.1. Design of beam current detector

Before developing a current transformer, several soft magnetic alloy materials were tested and compared. Finally, an iron-based soft magnetic alloy core was selected due to its high magnetic permeability (μ_r), low coercivity (H_c), high saturation magnetic induction (B_s) and low cost. In the design of the transformer, it is necessary to consider that the droop of the 500 μs macro pulse signal should be less than 1% within the pulse width, that is, $\tau_{droop} > 0.05$ s or $f_{low\text{-}cutoff} = 7$ Hz. According to the measured noise level of the experiment, the turns of the secondary coil N_s was selected as 50. An additional coil of 1 turn was wound for the online calibration. Since the inner diameter of the vacuum pipeline is Φ110 mm, a soft magnetic core of Fe-based nanocrystalline alloy (nominal relative magnetic permeability μ_r ≈90000@10 Hz) with an inner diameter D_i = 130 mm, outer diameter D_o = 160 mm and longitudinal length h = 25 mm was selected for winding the current transformer. The equivalent inductance L_s of the sensor was calculated according to equation (1) as 0.175 H. It was measured by an LCR meter (HIOKI IM3536) to be 0.176 H@10 Hz, which is close to the theoretical calculation.

$$L_S = N_S{}^2 \frac{\mu_0 \mu_r f_{fill}}{2\pi} h ln \frac{D_o}{D_i} \qquad (1)$$

Here in equation (1): $\mu_0 = 4\pi\times10^{-7}$ H/s is the vacuum magnetic permeability and f_{fill} = 0.75 is the filling factor of the ribbons forming the soft magnetic core. The input impedance of electronics r = 2.5

Ω, leads to the droop time constant of the pulse signal $\tau_{droop} = L_s/r = 0.062 > 0.05$ s, which meets the design requirements.

As shown in Fig. 1, the APEP current transformer is installed at the downstream of a bending magnet, which is used to guide the associated proton beam to the L-Dump or the APEP beam line. The stray magnetic field may cause the current transformer of APEP-CT to be magnetically saturated and fail to accurately measure the beam current strength of associated protons, so it is necessary to design a magnetic shield semi-surrounding the outside of the current transformer, as the inner shielding shown in Fig. 2. Considering the environmental stray magnetic field, a double-layered DT4 magnetic shield is designed according to equation (2). In this formula, $p = b^2/a^2$, where a and b are respectively the outer diameter and inner diameter of the cylindrical magnetic shield layer, and μ_r is the relevant permeability of magnetic material [4]. The magnetic shielding efficiency of a double-layered DT4 shield is calculated as a product of that of each layer with a radial thickness of 1 mm, that is, 37.8×38.7=1463, which is better than that of a single layer permalloy shield (S of which is 1220) with a radial thickness of 2.0 mm and a same outer diameter of Φ162 mm. The comparison of different magnetic shield is showed in Table 2. What's more, the material of DT4 is easy to process and costs much less. It can be seen that compared with a single-layer permalloy shield, this scheme has low cost and can meet the requirements of current transformers for enough magnetic shielding effect. A fully-enclosing outer shield of the APEP-CT sensor acts as a low-impedance path of the wall current and an RF shielding [5]. Figure 3 shows proton beam intensity detector installed at the APEP tunnel.

$$S = 4\mu_r p/(\mu_r^2 + 1)(p - 1) + 2\mu_r(p + 1) \quad (2)$$

Table 2. Calculation of Different Magnetic Shields as the Inner Shielding of APEP-CT

Layer Material	b (mm)	a (mm)	p	μ_r	S
DT4	162.0	160.0	1.0252	6000	37.8
DT4	158.0	156.0	1.0258	6000	38.7
Permalloy	162.0	158.0	1.0513	106	1220

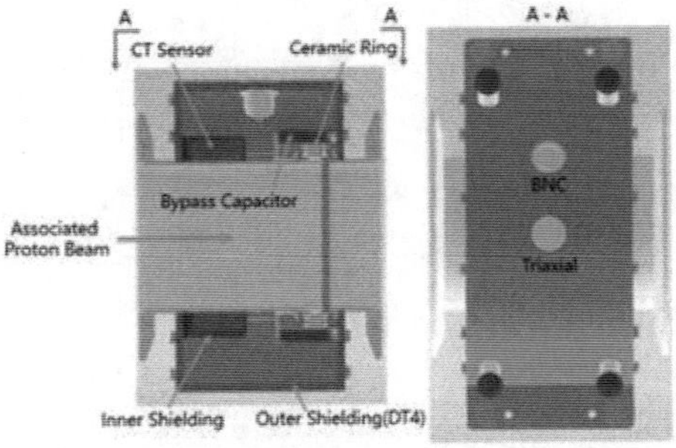

Figure 2. APEP CT sensor and magnetic shields.

Figure 3. Proton beam intensity detector installed at the APEP tunnel.

2.2. *Design of electronic circuit*

As a current transformer, when the sensor detects the proton beam current, the induced output signal is a very weak current (here N_s=50). In the design of electronics, we use a transimpedance amplifier as the first-stage IV conversion chip. For instance, a 10 μA beam current passes through the sensor core with the 50-turn secondary coil, the input current of the electronics will be 200 nA. The input and feedback resistances of the electronics are $r = 2.5$ Ω and $R_f = 10$ kΩ * 20 separately. When selecting a low-noise operational amplifier, the input bias current I_b cannot be ignored. The bandwidth of the electronics should be designed to minimize high frequency noise introduction while meeting the rise time requirement. The circuit diagram is shown in Fig. 4. The electronic baseline is automatically adjusted by a technique of baseline dynamic feedback [6].

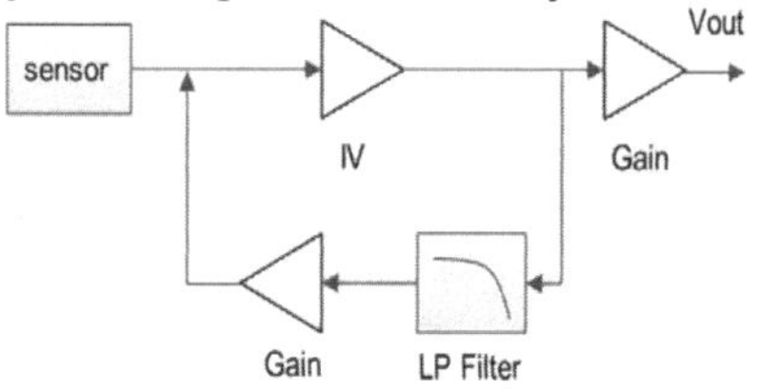

Figure 4. Diagram of CT electronics.

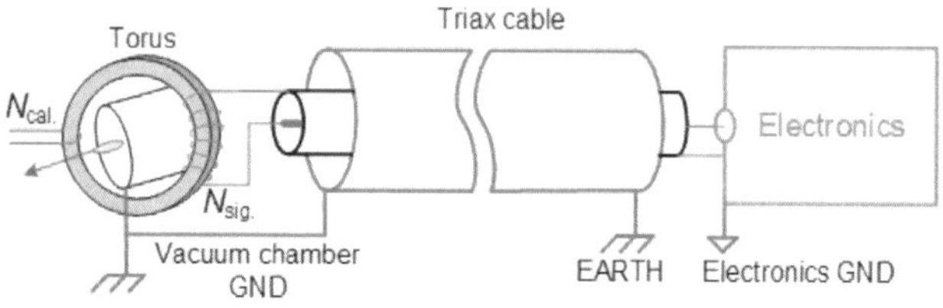

Figure 5. Diagram of the triax cable grounding to minimize noises.

3. System calibration and testing

3.1. System calibration

We used a high precision pulsed current source Keithley 6221 to generate a current of square wave with a 500 µs pulse width in a range of 2~10 µA, which passed through the calibration coil of the APEP-CT sensor to simulate the beam current. The induced current in the secondary coil is input to the electronics via a 25 m triaxial cable and the electronics output voltage is measured by an oscilloscope or JYTEK DAQ for the data acquisition and waveform display. The calibration formula is obtained by a linear fitting of the electronic output amplitude with the simulated input beam current intensity. Through this formula, the undertest beam intensity can be calculated along with the background denoising at a homometric beam-off zone during operation. Thereby we realized a precise low intensity measurement of the associated pulsed proton beam. After the system commissioning, the rise time of the electronics achieved < 10 µs, the droop is < 1 %/ms, and the relevant linear fitting error of the calibration data is within 1 % FS, as shown in Figs. 6 and 7. The parameters of this low intensity measurement system meet the design requirements.

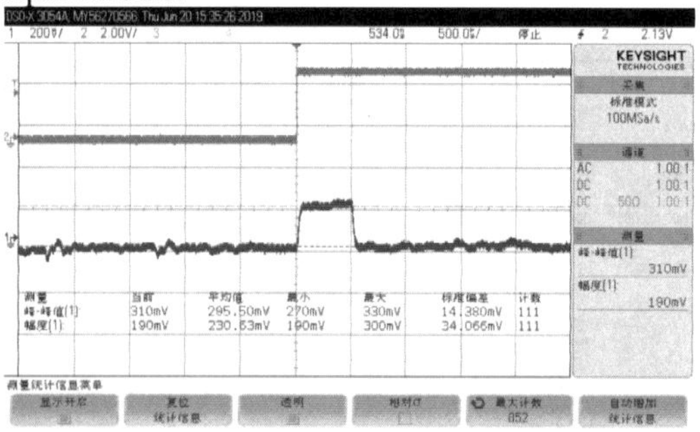

Figure 6. The blue signal is the electronics output with 1 µA calibrated input current.

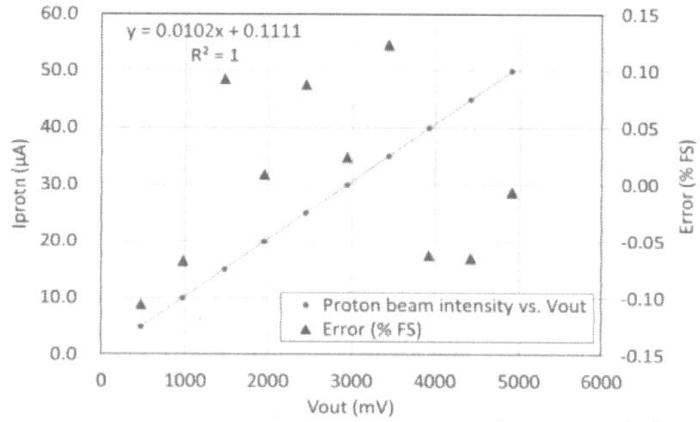

Figure 7. CT Electronics calibration data.

3.2. Troubleshooting and experimentation for interference investigation

The low beam intensity measurement system was installed in the APEP tunnel during the summer maintenance in 2021. In the following accelerator operation, we performed several beam tests with this system, which needs guiding the beam to the APEP beamline. It performed abnormal due to significant interferences. The interference was large and stable, characterized by a fixed frequency relating to the beam current. A Faraday cup at the end of the APEP beam line detects the same interference signal, but with a relatively smaller impact, shown in Fig. 8.

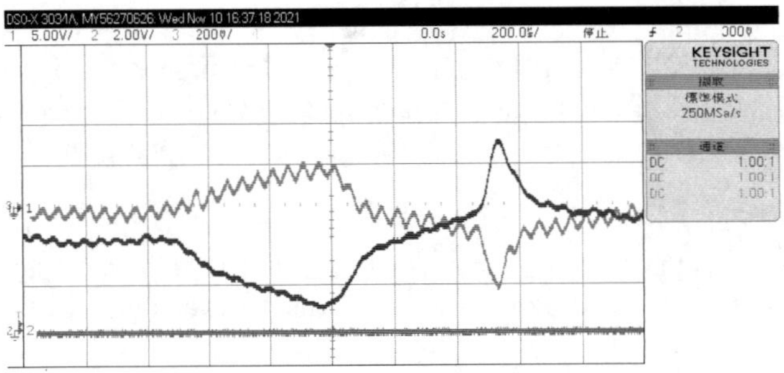

Figure 8. Measurement of the APEP-CT and Faraday cup electronic output waveforms. Blue trace is the signal of Faraday Cup. Yellow trace is that of APEP-CT. Purple trace is the timing trigger signal.

We made a thorough examination of the CT interference sources along the APEP beamline and RCS. The magnetic elements such as Linac RCS Switch Magnet, Painting Bump Magnet at the RCS injection zone, and Extraction Magnets at the extraction zone were sequentially shut down. The position of the magnet is shown in Fig. 1. The BH (Horizontal painting bump magnets) and BV (vertical painting bump magnets) are used for the horizontal and vertical painting respectively in the phase space of the injected beam. The main contribution of interference is from BH, followed by BV, which has a high similarity to the current curve of the power supply during the beam injection in the phase space painting mode. The correlation indicated that the interference was transmitted from the orbit bumper to the APEP-CT sensor along the accelerator vacuum pipe. The distance is about 55 meters. While other interference effects to the APEP-CT are small enough to be ignored.
Due to the consistent shape and amplitude of the interference signal, the DAQ system is triggered by the Timing system, allowing it to collect the interference signal once without a beam and save it as the background interference signal. Subsequently, for each subsequent signal collection with a beam, the interference signal is subtracted in order to restore the beam signal.

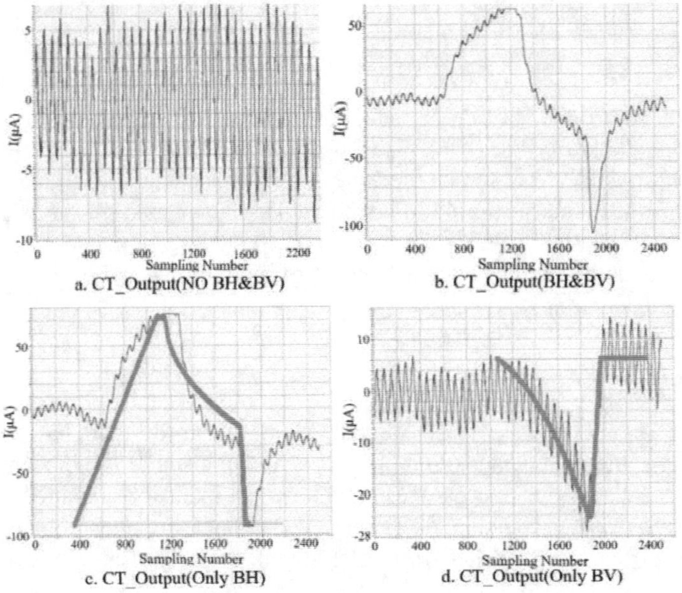

Figure 9. Output of APEP-CT electronics under different accelerator operation conditions. Pic (a) without BH and BV magnets operation. Pic (b) with the BH & BV magnet enabled. Pic (c) with only the BH magnet enabled. Pic (d) with only the BV magnet enabled.

To keep the interferences and high harmonics of the beam out of the sensor cavity, a bypass capacitor was designed (as shown in Fig. 2), and installed during the summer maintenance in 2022. The design principle of the bypass capacitor is described in Ref. [7]. The wall current splits in two: the high frequencies pass through the capacitance of the ceramic gap, which cancel the high frequency part of the beam “seen” by the sensor, and the low frequencies follow the wall current bypass, therefore do not pass through the sensor hole. Here, we built a capacitor over the ceramic gap with layers of copper foil separated by a layer of 100 μm-thick Kapton foil. The cavity impedance was dominated by the nanocrystalline core, which is measured to be 63.8 Ω@1 kHz and will be much bigger in higher frequencies. The bypass capacitance is designed as 5 nF, allowing the main part (<50 kHz) of the beam passing through the sensor hole. To obtain the desired capacitance value, the overlapping area is obtained by equation (3)[8]:

$$S = C\, d/\varepsilon_r \varepsilon_0 \tag{3}$$

where: C is the capacitance [F], $S = 1.61\times10^{-2}$ is the overlapping area [m^2], $d=1\times10^{-4}$ is the dielectric thickness [m], ε_r is the relative dielectric constant, 3.5 for Kapton polyimide, ε_0 is the vacuum dielectric constant 8.86 x 10^{-12} F/m.

Figure 10. A bypass capacitor over the ceramic gap was built in the APEP-CT sensor.

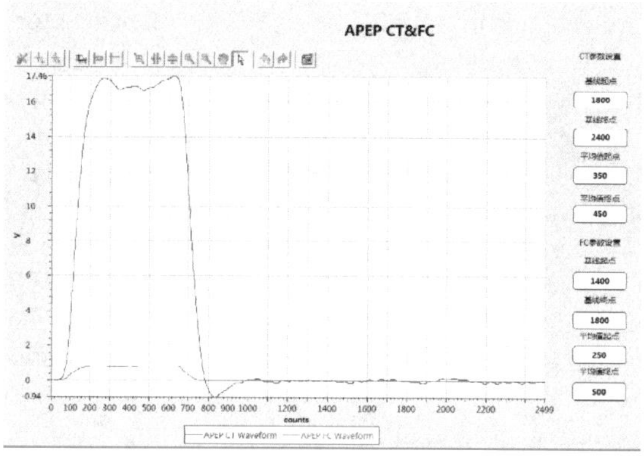

Figure 11. Output of the electronics of the APEP-CT and Faraday cup.

3.3. *Test results*

After the bypass capacitor was installed in the APEP-CT sensor, we successfully measured the associated proton beam in low intensity, shown in Fig. 11. It was observed that the interference from the BH and BV magnets diminished significantly, and in fact, became barely noticeable. The signal-to-noise ratio of the beam signal was improved, and the amplitude of the pulsed associated proton beam intensity is clearly measured as 16.8 μA. The blue waveform represents the signal from the APEP-CT system, while the red waveform represents the current intensity signal measured by the Faraday cup after the beam passes through the energy degrader.

4. Conclusions

In the beam line of the Associated Proton beam Experiment Platform at CSNS, the proton beam intensity is a crucial parameter for irradiation experiments. The APEP-CT system achieves a signal-to-noise ratio improvement and meets the design specification for microampere-level current intensity measurement by implementing methods such as, the bypass capacitance design, the cable shielding and grounding methods, a low-noise electronic with the correct LP-filter design, and background noise subtraction algorithms in DAQ.

References

[1] Wang S, An Y and Fang S 2011 *Sci. China Phys. Mech. Astron* An overview of design for CSNS/RCS and beam transport **54** pp 239–44

[2] Xu S and Liu H 2019 *Proc. IPAC'19* Beam Commissioning Experience of CSNS/RCS vol 15 (Australia: Melbourne) pp 1012-14

[3] Liu Yingyi and Jing Hantao 2022 *Nucl. Instrum. Meth. A* Physical design of the APEP beam line

at CSNS vol 1042 pp 167431
[4] Feng G and He Z 2010 *Acta Metrologica Sin* Calculation and simulation of geomagnetic field shielding effectiveness of projectile body **31** pp 141-44.
[5] Bergoz Instrumentation 2021 AC Current Transformer User’s Manual Rev.5.4
[6] Zhou Q and Wang Z 2005 *Proc. SPIE’06* A DC-servo baseline restorer and its implementation in biomedical instrumentation vol 6040 pp 60401K
[7] Huang W L and Qiu R Y 2019 *Proc. IBIC'19* A Dual Functional Current Monitor for Stripping Efficiency Measurement in CSNS (Sweden: Malmö) pp 96-99
[8] Bergoz Instrumentation 2014 New Parametric Current Transformer User's Manual Rev.1.1

Multi-tile zinc-oxide-based radiation-hard fast scintillation counter for relativistic heavy-ion beam diagnostics: prototype design and test

M Saifulin[1,2], P Boutachkov[1], E Gorokhova[3], P Rodnyi[4], C Trautmann[1,2], I Venevtsev[4] and B Walasek-Höhne[1]

[1] GSI Helmholtz Center for Heavy Ion Research GmbH, Planckstrasse 1, Darmstadt 64291, Germany
[2] Technical University Darmstadt, Karolinenplatz 5, Darmstadt 64289, Germany
[3] Joint Stock Company "Research and Production Corporation S.I. Vavilova," Babushkina street 36/1, St. Petersburg 192171, Russia
[4] Peter the Great Polytechnic University, Polytechnicheskaya 29, St. Petersburg 195251, Russia

E-mail: M.Saifulin@gsi.de

Abstract. This contribution summarizes the design and performance test of a prototype radiation-hard fast scintillation detector based on the indium-doped zinc oxide ceramic scintillator, ZnO(In). The prototype detector has been developed for use as a beam diagnostics tool for high-energy beam lines of the SIS18 synchrotron at the GSI Helmholtz Center for Heavy Ion Research GmbH. The new detector consists of multiple ZnO(In) scintillating ceramics tiles stacked on the front and back sides of a borosilicate light guide. The performance of the detector was tested in comparison to a standard plastic scintillation detector with 300 MeV/u energy ^{40}Ar, ^{197}Au, ^{208}Pb, and ^{238}U ion beams. The investigated prototype exhibits 100% counting efficiency and radiation hardness of a few orders of magnitude higher than the standard plastic scintillation counter. Therefore, it provides an improved beam diagnostics tool for relativistic heavy-ion beam measurements.

1. Introduction

The heavy-ion accelerator facility at the GSI Helmholtz Center for Heavy Ion Research (Darmstadt, Germany) and other similar ion-beam facilities worldwide provide energetic heavy-ion beams for a wide range of research purposes, including fundamental nuclear and particle physics, atomic and plasma physics, material sciences, radiation biology, and cancer therapy research.

Various diagnostics systems are used to measure heavy-ion beam parameters to ensure the safety and reliability of the accelerator facility operation, as well as to optimize the beam delivery to the experimental locations. For example, this includes beam position, profile, intensity, energy, and emittance measurements.

The information provided by the beam diagnostics tools is used by the operating team for the online feedback controls of the accelerator and beam transfer lines. This helps to maintain the desired quality and stability of the beam. The beam diagnostics systems are also used by experimental users directly for the characterization and calibration of the experimental setups, which require precise knowledge of the incoming beam parameters.

Journal of Physics: Conference Series **2681** (2024) 012002 doi:10.1088/1742-6596/2681/1/012002

In particular, at the high-energy beam transfer lines (HEST) of the GSI facility, the intensity of heavy-ion beams slowly extracted from the SIS18 synchrotron is measured using three different detector types (particle detector combination, PDC): scintillation counter (SC), ionization chamber (IC) and secondary electron monitor (SEM). Each detector type is used to cover a specific range of beam intensity [1].

Plastic scintillation counters based on BC-400 (Saint Gobain), or EJ-212 (Eljen Technology) scintillators are currently used as a part of the HEST ion beam intensity diagnostics, covering counting rates in the range up to 1×10^6 ions/s. The heavy-ion beams induce a large amount of radiation damage when passing through the plastic scintillator. This leads to frequent detector services where plastic scintillators are exchanged, which is highly undesirable from an operational point of view.

In this work, we investigated an alternative to the plastic scintillation counter, that could provide less frequent detector maintenance through the use of a more radiation-hard scintillator. We developed a new detector prototype based on indium-doped zinc oxide (ZnO(In)) ceramic scintillator. We report on the new prototype design and performance, which was evaluated experimentally through in-beam tests and simulations using the OpenGATE [2].

2. Zinc oxide ceramic scintillator

Zinc oxide (ZnO) is a well-known inorganic compound used in various applications [3, 4, 5]. In particular, it has been used for the detection of X-rays and α-particles [6, 7, 8, 9].

ZnO exhibits two luminescence bands when excited by light or ionizing radiation: (1) emission in the ultraviolet (UV) spectral range ($\sim$390 nm) with a short scintillation decay time ($<$1 ns), and (2) emission in a broad-band with a maximum around 550 nm wavelength (green luminescence) with a longer decay time ($>$1 s). The green luminescence is unsuitable for fast-counting applications. Therefore, it is avoided either by annealing in a reducing environment or doping with a group 3A element impurities, such as Al, Ga, or In [3, 10, 11].

In-doped and Ga-doped ZnO ceramics prepared by uni-axial hot pressing in vacuum have been investigated as promising scintillators for fast heavy ion counting applications [12, 13, 14, 15]. They exhibit orders of magnitude higher radiation hardness than plastic scintillators, and thus can provide longer exploitation time before scintillator replacement is required. Furthermore, narrow signals with the pulse width below 1 ns can be obtained, when ZnO(In) scintillation is registered with a fast photomultiplier tube. This resolves pile-up limitations for the micro-spill structure measurements [13].

The fast scintillation component in ZnO is located close to its self-absorption edge. Thus, the detection of fast scintillation photons through a few millimeters thick ZnO is not efficient. This peculiarity and difference compared to standard plastic scintillators have been considered in the new ZnO-based prototype detector design.

3. Prototype detector details

3.1. Prototype detector's parts

The detector consists of three main components: the active volume, the light guide, and the photosensor, as shown in Fig. 1.

The active volume is the central part of the detector that directly interacts with the incident ion beam. It consists of square-shaped scintillator tiles made of ZnO(In), stacked on both the front and back sides of a borosilicate glass plate. When the ion beam passes through the active volume, it deposits part of its energy in the scintillator and borosilicate glass layers, leading to the emission of scintillation light.

The scintillation mainly comes from ZnO(In) layers, as they efficiently convert the deposited energy into light photons. Due to self-absorption, the scintillation photons mainly go out from the material through the surfaces where ions enter and exit the scintillating tile. The

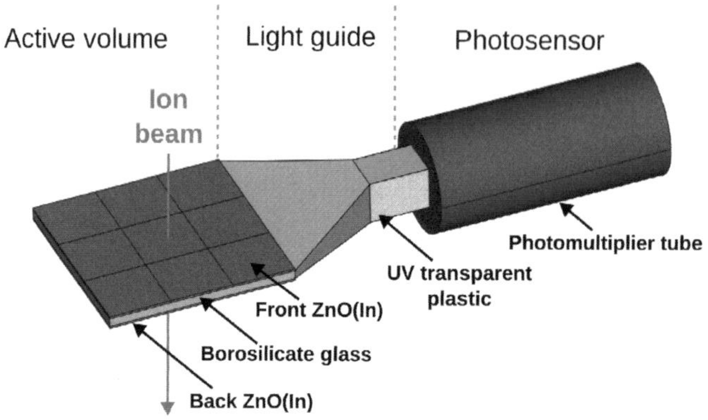

Figure 1. Schematic view of the multi-tile ZnO-based scintillation counter prototype.

borosilicate glass, which is transparent to scintillation photons from ZnO(In), acts as a light guide, assisting in collecting scintillation light from the ZnO(In) surfaces. This arrangement enables the extraction of scintillation photons from the active volume.

An additional light guide made of UV-transparent plastic is used to match the shapes of the borosilicate glass and photosensor. Optical silicon (Momentive, RTV-615) improves light coupling between the plastic light guide and both borosilicate glass output and photosensor input.

The photosensor, responsible for converting scintillation light into an electrical output signal, consists of a photomultiplier tube (PMT, Hamamatsu R6427).

The active volume and the light guide are fixed on a 3D-printed holding structure and wrapped with Teflon to improve light collection and black tape to avoid penetration of any external light.

In the design of the new ZnO-based scintillation counter, only the active volume component is replaced, while the plastic light guide and photomultiplier tube remain unchanged compared to the standard plastic scintillation counter for GSI beam diagnostics. This design approach ensures the proven performance and compatibility with existing plastic scintillation counter components, while introducing the enhanced scintillation properties of ZnO(In) in the active volume.

3.2. Materials of the active volume

The 0.046 at.% In^{3+} doped ZnO ceramic scintillators have been produced at the Joint Stock Company "Research and Production Corporation S.I. Vavilova" (St. Petersburg, Russia) using the uni-axial hot pressing in vacuum. The obtained ZnO(In) ceramics were polished and cut into $15 \times 15 \times 0.5\,\mathrm{mm}^3$ square tiles.

The borosilicate radiation-hard glass (material catalog name BK7G18) was purchased from SCHOTT (Germany) and cut into the $50 \times 50 \times 2\,\mathrm{mm}^3$ rectangular shape. The material surfaces had a root-mean-square roughness of 2–4 m.

The ZnO(In) ceramic tiles and BK7G18 plate are fixed together using a Kapton tape as shown in Fig. 2.

4. Experimental performance test

To evaluate the performance of the prototype detector, a series of tests were conducted at GSI using ion beams extracted from the SIS18 accelerator. The experimental parameters are summarized in Table 1.

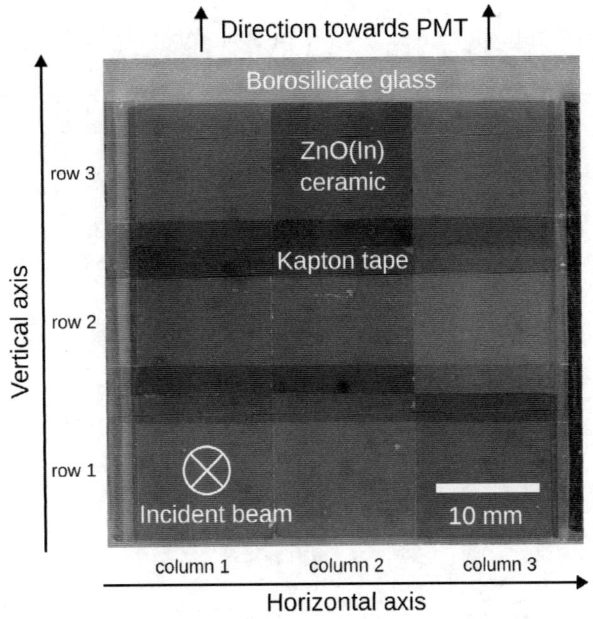

Figure 2. Photograph of the prototype detector's active area. The white crossed circle shows the incident beam direction and the beam spot size in scale with the active area.

The ^{40}Ar, ^{197}Au, ^{208}Pb, and ^{238}U ions had an initial energy of 300 MeV/u. Before reaching the entrance of the prototype detector, the ion beams underwent energy loss. This loss occurs as the beam exits the vacuum chamber through a 100 m thick stainless-steel window and travels a 40 cm distance through air. To accurately determine the incident beam energies and quantify the energy losses within the detector's active volume layers, the ATIMA code was employed [16]. For the energy loss calculations, we used 5.5 g/cm^3 ZnO(In) and 2.58 g/cm^3 BK7G18 material density.

The ion beam was collimated to a circular spot with a diameter of 5 mm before interacting with the prototype detector. The prototype detector was mounted on a stage capable of horizontal and vertical movement, enabling exposure of different regions of the active volume to the ion beam.

A standard plastic scintillation counter was placed behind the prototype and used as the reference detector.

Output signals from both the prototype and reference detectors were digitized and read out with a high-speed oscilloscope (Tektronix MSO58) with a bandwidth of 2 GHz. The reference detector provided a trigger for the oscilloscope read-out. The counting efficiency of the prototype detector relative to the reference detector was estimated by counting the number of prototype detector signals higher than 50 mV.

The PMT gain of the prototype detector was adjusted so that the maximum amplitude of the signal is of the order of 1 V.

5. Results and discussion

5.1. Contribution of borosilicate glass signal

By scanning the incident ion beam near the edge of the prototype's active volume, we collected signals corresponding to ions passing through: (1) only borosilicate glass plate, (2) borosilicate

Table 1. Experimental beam parameters and photomultiplier gains used for the tests of the prototype detector.

Ion	Energy, MeV/u	Energy loss, MeV	PMT gain
^{40}Ar	300 ^{a}	231.85 ^{c}	2.20×10^6
	297.14 ^{b}	516.89 ^{d}	(1300 V)
		240.39 ^{e}	
^{197}Au	300 ^{a}	4654.27 ^{c}	1.15×10^5
	288.62 ^{b}	10900 ^{d}	(850 V)
		5499.17 ^{e}	
^{208}Pb	300 ^{a}	5006.94 ^{c}	1.15×10^5
	288.41 ^{b}	11800 ^{d}	(850 V)
		5932.97 ^{e}	
^{238}U	300 ^{a}	6272.86 ^{c}	1.15×10^5
	287.32 ^{b}	14900 ^{d}	(850 V)
		7586.86 ^{e}	

^{a} SIS18 extraction energy, ^{b} Energy at the prototype detector entrance, ^{c} front ZnO(In), ^{d} borosilicate glass, ^{e} back ZnO(In)

glass with one layer of ZnO(In), and (3) borosilicate glass with two layers of ZnO(In).

As shown in Fig. 3, the scintillation light yield of the borosilicate glass (BSG) is lower compared to the ZnO(In) scintillator. In the case of Au ion irradiation, the signal amplitude from the borosilicate glass layer is around 30 mV, while the signal amplitude from borosilicate glass with one layer of ZnO(In) is almost 3 times larger. The signal from the borosilicate glass with two layers of ZnO(In) scintillator is roughly 5 times larger than the signal from the borosilicate glass.

Despite the lower energy loss in ZnO(In) compared to borosilicate glass (see Table 1), ZnO(In) scintillation light represents the main part of the output signal amplitude, indicating that it is a brighter scintillator than the BK7G18.

For all three investigated regions, the registered output signals have a rise time of 2.5 ns. This rise time is related to the response of the PMT and has nothing to do with the scintillation material and beam position in the active area.

The use of borosilicate glass provides a solution to overcome the challenge related to the self-absorption of ZnO(In). It provides efficient extraction of scintillation light out from the active volume and, at the same time, it has a negligible contribution to the prototype's output signal.

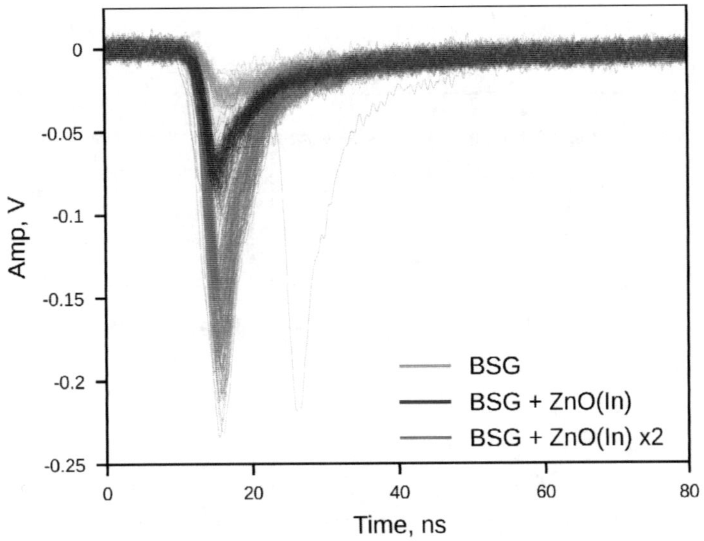

Figure 3. Signals of the prototype detector exposed to 300 MeV/u Au ion beam at the edge of the active volume.

5.2. Prototype detector signals across active volume

Figure 4 shows the prototype output signals registered when different ion beams hit the active volume at different locations. We collected 10000 output signals corresponding to individual ions passing through the center of each ZnO(In) scintillating tile located at various horizontal and vertical positions (various columns and rows), as is shown in Fig. 2.

Depending on the beam position, there is no difference in the signals registered along the horizontal axis (fixed row, various columns). In contrary to that, there is a significant change in the output signal amplitude along the vertical axis (fixed column, different rows). The signal amplitude dependency along the vertical axis has a similar trend for all the investigated ion species.

It should be noted that the plastic light guide and PMT are placed along the vertical axis. The higher amplitude signals are registered when the beam hits the active volume closer to the PMT (row 3 in Fig. 4).

5.3. Light propagation simulation

The propagation of the scintillation light from ZnO(In) through BK7G18 glass was simulated using the OpenGATE [2]. As a result of the simulation, it was found that the ZnO(In) scintillation photons transmission through the BK7G18 glass varies from 2 to 9% depending on the vertical axis coordinate. Figure 5 shows a comparison of the experimentally measured signal amplitude and the simulation result, confirming that the prototype signal amplitude variation is consistent with the ZnO(In) scintillation light transmission through the BK7G18 glass.

Further borosilicate glass plate optimizations (geometry and surface quality) together with OpenGATE simulations are required to find the optimal shape and surface quality of the borosilicate glass to suppress the changes in the output signal amplitude along the vertical axis of the active volume.

5.4. Prototype detector counting efficiency

The counting efficiency has been estimated as a ratio of the number of ions counted by the prototype divided by the number of ions counted by the reference detector. Despite the signal

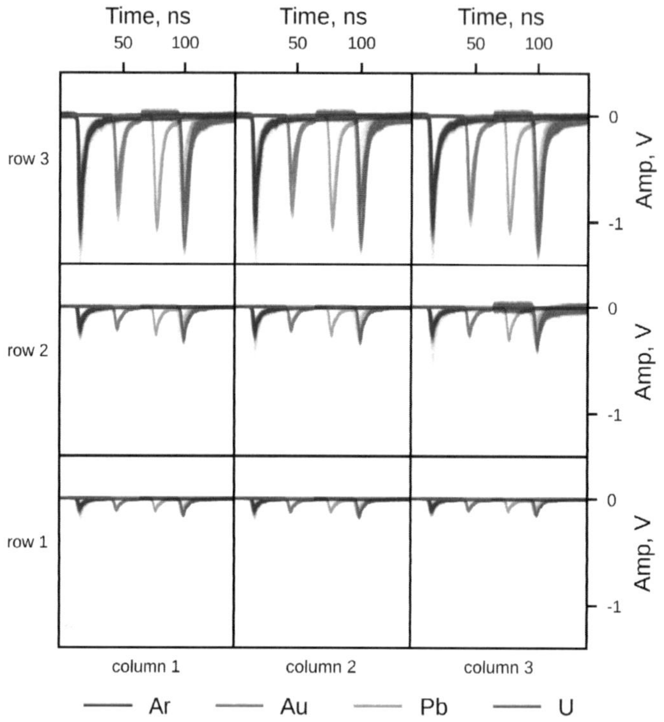

Figure 4. Output signals obtained for various ion beams hitting the prototype detector at different horizontal and vertical positions. Signals of different ion species are shifted on the x-axis for better visibility and comparison. Each subplot corresponds to the center of the scintillating tile in the corresponding column and row, as shown in Fig. 2.

amplitude variation in the vertical axis, the prototype detector exhibits 100% counting efficiency relative to the plastic scintillator for all the investigated ion species.

6. Summary

We developed a new radiation-hard fast scintillation detector prototype based on ZnO(In) transparent scintillation ceramics to be used for relativistic heavy-ion beam intensity diagnostics.

The developed prototype showed 100% counting efficiency relative to the standard plastic scintillation counter when tested with Ar, Au, Pb, and U heavy-ion beams at 300 MeV/u energy. By considering the few orders of magnitude higher radiation hardness of ZnO(In) than plastic scintillator, the new prototype provides an improved beam diagnostics tool.

We observed a spatial dependency of the prototype output signal amplitudes registered when the ion beam hits different positions on the active volume. We showed that this dependency is related to the ZnO(In) scintillation light transmission through the borosilicate glass. The experimentally observed signal amplitude change is in agreement with the light transmission through the borosilicate glass simulated by OpenGATE.

The borosilicate glass, used as a part of the active volume, exhibited scintillation light yield much lower than ZnO(In). Thus, it makes a negligible contribution to the prototype detector output signal and serves mainly as the light guide.

In the current prototype version, we used a rectangular-shaped borosilicate light guide. Further investigations of more complex shapes of the light guide are needed to eliminate the

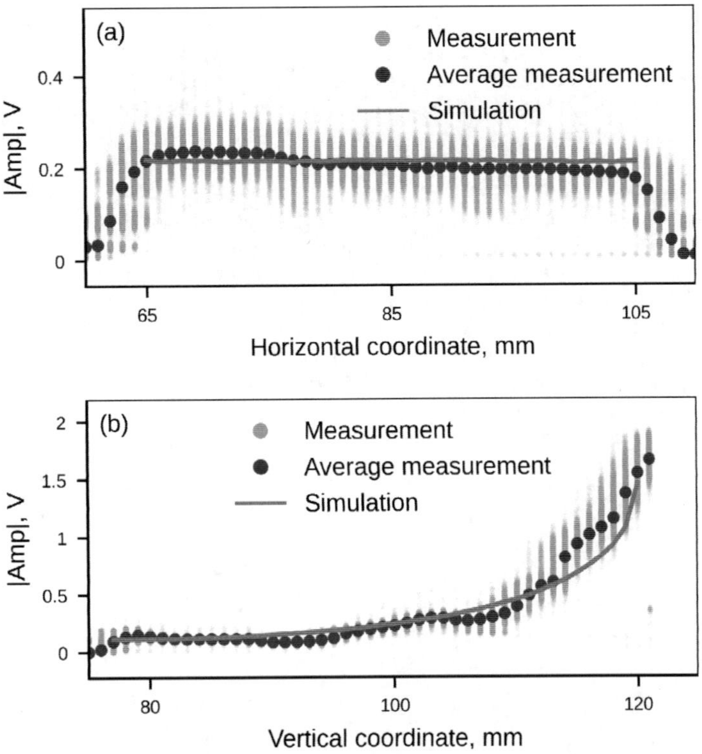

Figure 5. The signal amplitude changes for the 300 MeV/u Au beam incident at various positions on the active volume. Figures (a) and (b) correspond to beam scans along the horizontal and vertical axis, respectively.

spatial dependency of the output signal.

Acknowledgments
DLR funded this work within the ERA.Net RUS Plus Project RUS_ST2017-051.

Special consideration
The results presented in this contribution are based on the work performed before the 24th of February 2022.

References
[1] Forck P, Heeg P and Peters A 1997 Intensity measurement of high-energy heavy ions at the GSI facility *AIP Conf. Proc.* vol **390** (American Institute of Physics) pp 422–429
[2] Jan S, Santin G, Strul D, Staelens S, Assié K, Autret D, Avner S, Barbier R, Bardies M, Bloomfield P *et al.* 2004 *Phys. Med. Biol.* **49** 4543
[3] Özgür Ü, Alivov Y I, Liu C, Teke A, Reshchikov M A, Doğan S, Avrutin V, Cho S J and Morkoç H 2005 *J. Appl. Phys.* **98** 041301
[4] Klingshirn C, Fallert J, Zhou H, Sartor J, Thiele C, Maier-Flaig F, Schneider D and Kalt H 2010 *Phys. Status Solidi B* **247** pp 1424–1447
[5] Sahani R and Dixit A 2022 *Mater. Sci. Semicond. Process.* **151** 107040
[6] Zhang Q, Yan J, Deng B, Zhang J, Lv J, Wen X and Gao K 2017 *J. Instrum.* **12** P12033
[7] Neal J S, Boatner L A, Giles N, Halliburton L, Derenzo S and Bourret-Courchesne E 2006 *Nucl. Instrum. Methods Phys. Res., Sect. A* **568** pp 803–809

[8] Neal J S, DeVito D M, Armstrong B L, Hong M, Kesanli B, Yang X, Giles N C, Howe J Y, Ramey J O, Wisniewski D J *et al.* 2009 *IEEE Trans. Nucl. Sci.* **56** pp 892–898
[9] Sahani R, Kumari C, Pandya A and Dixit A 2019 *Sci. Rep.* **9** 11354
[10] Gorokhova E, Eron'ko S, Kul'kov A, Oreshchenko E, Simonova K, Chernenko K, Venevtsev I, Rodnyĭ P, Lott K and Wieczorek H 2015 *J. Opt. Technol.* **82** pp 837–842
[11] Gorokhova E, Eron'ko S, Oreshchenko E, Sandulenko A, Rodnyĭ P, Chernenko K, Venevtsev I, Kul'kov A, Muktepavela F and Boutachkov P 2018 *J. Opt. Technol.* **85** pp 729–737
[12] Boutachkov P *et al.* 2019 Radiation hardness investigation of zinc oxide fast scintillators with relativistic heavy ion beams *Proc. IBIC'19* (JACoW Publishing, Geneva, Switzerland) pp 71–73
[13] Boutachkov P *et al.* 2022 Novel fast radiation-hard scintillation detectors for ion beam diagnostics *Proc. IBIC'22* (JACoW Publishing, Geneva, Switzerland) pp 515–518
[14] Saifulin M *et al.* 2022 ZnO(In) scintillation light spectra investigation for heavy ion detector application *Proc. IBIC'22* (JACoW Publishing, Geneva, Switzerland) pp 294–295
[15] Saifulin M, Boutachkov P, Gorokhova E, Rodnyi P, Simon P, Trautmann C, Venevtsev I and Walasek-Höhne B 2022 *J. Appl. Phys.* **132** 195901
[16] Geissel H, Weick H, Scheidenberger C, Bimbot R and Gardes D 2002 *Nucl. Instrum. Methods Phys. Res., Sect. B* **195** pp 3–54

Journal of Physics: Conference Series **2681** (2024) 012003 doi:10.1088/1742-6596/2681/1/012003

A novel BPM mechanical center calibration method based on laser ranging

X H Tang[1,2], Y F Sui[1,2], J He[1], Y Y Du[1], J S Cao[1] and J H Yue[1*]

[1]Institute of High Energy Physics, Chinese Academy of Science, Beijing 100049, China

[2]School of Nuclear Science and Technology, University of Chinese Academy of Science, Beijing 100049, China

E-mail: yuejh@ihep.ac.cn

Abstract. Determining the mechanical center of the beam position monitor (BPM) has been a difficulty for BPM calibration. To solve this problem, a method of positioning the BPM mechanical center based on laser ranging is proposed. This method uses high-precision antenna support as the core locating datum, and high-precision laser ranging sensor (LRS) as the detection tool. By detecting the distances from the LRSs to the antenna support and the distances from the LRSs to the BPM, the mechanical center of the BPM can be indirectly determined. The theoretical system error of this method is within 20 μm, and the experimental results show that the measurement repeatability is 15 μm, This method has low cost and fast speed, which can be used for large-scale calibration.

1. Introduction

The BPM system, as the eyes of the particle accelerator, plays an important role in the stability of the beam orbit. About 600 BPMs are produced during the construction of High Energy Photon Source (HEPS) project [1]. Due to processing errors, the mechanical center and electrical center of BPM do not coincide. Therefore, each BPM is demanded to be calibrated before use [2]. For the button-type BPM with 45-degree rotation, as shown in figure 1, the relationship between position coordinates and electrode signal amplitudes is defined as Eq. (1) [3].

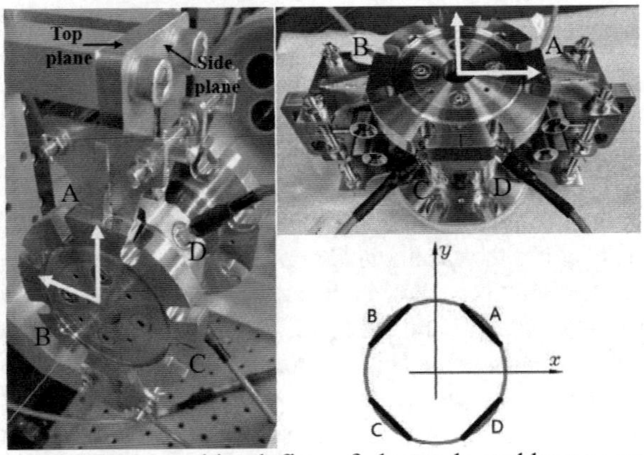

Figure 1. A HEPS BPM and its define of electrode and beam coordinate system.

$$x = K_x \frac{V_a + V_d - V_b - V_c}{V_a + V_b + V_c + V_d} + X_{\text{offset}} = K_x U + X_{\text{offset}}$$
$$y = K_y \frac{V_a + V_b - V_c - V_d}{V_a + V_b + V_c + V_d} + Y_{\text{offset}} = K_y V + Y_{\text{offset}} \tag{1}$$

where K_x, K_y are the BPM sensitivity coefficients, and X_{offset}, Y_{offset} are the difference between the BPM electrical center and mechanical center. The process of determining K_x, K_y, X_{offset}, and Y_{offset} is called BPM calibration. A BPM automatic calibration system is shown in figure 2 and it is composed of an RF signal source, antenna (Goubau line), precision motion stages and their controller, BPM electronics, and the industrial personal computer [4, 5]. The Goubau line emits transverse electromagnetic (TEM) waves with the excitation of the RF signal source, so as to simulate the electromagnetic field of the charged particle beam in the accelerator to enable BPM calibration [5]. The BPM is driven by the precision motion stages to move in horizontal and vertical directions. The industrial computer saves the real BPM position data recorded by the controller and the calculated BPM position data from the BPM electronics. Then, the sensitivity coefficients and offsets are analyzed by the software algorithm.

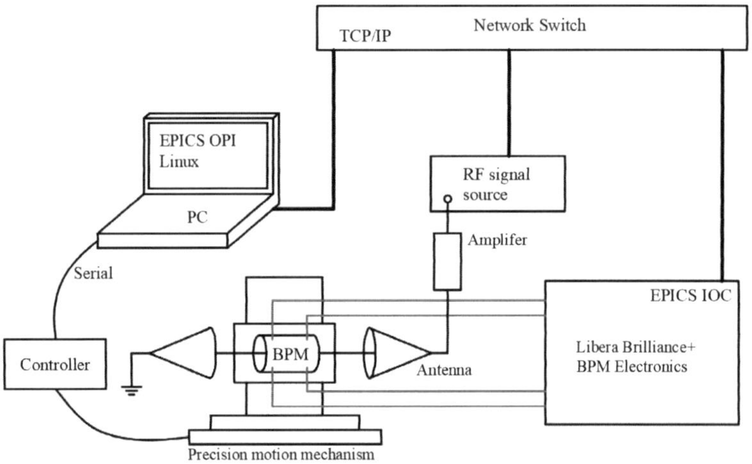

Figure 2. Schematic of BPM automatic calibration system.

In general, it is relatively easy to ascertain the BPM electrical center, as long as the operator observes the four channels of BPM electronics and makes them equal by adjusting the BPM motion stage. To achieve this goal, it is necessary to ensure that the four channels of BPM electronics and the coaxial cables are calibrated. However, finding the mechanical center is extremely difficult, so it is hard to determine X_{offset} and Y_{offset}.

2. A new approach to determining BPM mechanical center

The main difficulties in calibrating the mechanical center are as follows: firstly, it is not easy to ensure that the antenna is parallel to the axis of BPM; Secondly, it is almost impossible to directly measure the distance between the antenna and the edge of the BPM vacuum chamber; Finally, the high accuracy requirement (tens of micrometers) poses great challenges for manual adjustment. In order to solve the problems above, a high-precision antenna positioning support was designed, and a high-performance laser ranging sensor (span 30±4 mm, resolution 0.5 μm) was introduced as the detection tool for the system self-inspection and seeking BPM mechanical center. The schematic diagram of the positioning system is shown in figure 3, where the uppercase letters XYZ represent the geodetic coordinate system (GCS) and lowercase letters *xy* represent the BPM coordinate system. Figure 4 shows the distances that need to be measured to determine the BPM mechanical center, where H and L are the dimensions of BPM, H_s are the dimensions of the antenna support, r is the radius of the antenna, X_s and Y_s are the distances from the LRSs to the antenna support, and X_{bpm} and Y_{bpm} are the distances from the LRSs to BPM. When the antenna is located in the BPM mechanical center, the following relationship is satisfied:

$$\begin{aligned} Y_s + H_s &= Y_{bpm} + \frac{H}{2} + r \\ X_s &= X_{bpm} + \frac{L}{2} + r \end{aligned} \tag{2}$$

Transfer Eq. (2) to Eq. (3). When the position of BPM, namely X_{bpm} and Y_{bpm}, have been adjusted to meet Eq. (3), it can be considered that the mechanical center of BPM has been found.

$$\begin{aligned} Y_{bpm\_M} &= Y_s + H_s - \frac{H}{2} - r \\ X_{bpm\_M} &= X_s - \frac{L}{2} - r \end{aligned} \tag{3}$$

In Eq. (3), the right side of the equals sign is the parameters of the positioning system. Moreover, only Y_s and X_s are measured values, the rest are all constants. As long as the system state is unchanged, X_{bpm_M} and Y_{bpm_M} are two constants. When the antenna is in the electrical center of BPM, let the distance measured by the two laser sensors be denoted by X_{bpm_E} and Y_{bpm_E}, then the offset between the electrical center and the mechanical center can be calculated as:

$$\begin{aligned} X_{\text{offset}} &= -(Y_{bpm\_E} - Y_{bpm\_M}) \\ Y_{\text{offset}} &= X_{bpm\_E} - X_{bpm\_M} \end{aligned} \tag{4}$$

and the specific calibration operating steps are shown in table 1.

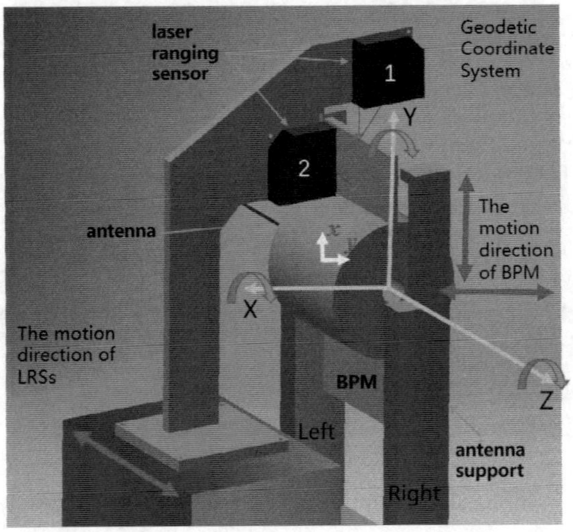

Figure 3. BPM machine center calibration method based on laser ranging.

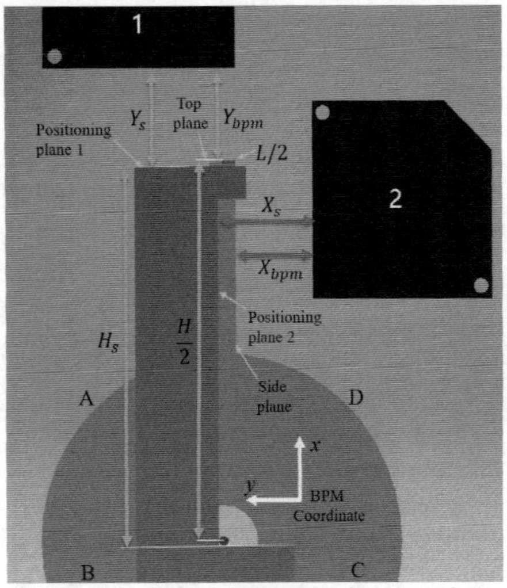

Figure 4. The distances that need to be measured to determine mechanical center.

In table 1, step 1 is a system self-test, which only needs to be performed once as long as the system state remains unchanged. Step 2 is to make the antenna parallel to the system, and steps 3-4 ensure that the BPM is parallel to the system. Steps 5 and 6 are optional operations. If calibrating from the mechanical center, proceed to step 5. If only measuring offset, perform step 6.

3. Error analysis

The processing errors of the antenna support and BPM is the main source of system errors. Now assume that the size errors of the support and BPM are ΔH_s and ΔH, so the measured value Y_s in Eq. (3) will turn into $Y_s + \Delta H_s$, and Y direction is $X'_{\text{offset}} = X_{\text{offset}} + \Delta H_s - \Delta H$. Similarly, $Y'_{\text{offset}} = Y_{\text{offset}} + \Delta L - \Delta F$, where ΔL is the size error of the BPM in the X direction, and ΔF is the geometry error of the support's positioning plane. The error model is shown in figure 5.

Considering the worst-case scenario, when the error direction of the antenna support is opposite to the BPM, the total systematic errors are the sum of absolute values:

$$\Delta X^{s}_{\text{offset}} = |\Delta H_s| + |\Delta H|$$
$$\Delta Y^{s}_{\text{offset}} = |\Delta L| + |\Delta F| \tag{5}$$

Since the machining error of both antenna and BPM is within 10 μm, the maximum system error is less than 20 μm.

The random error of this method includes the measurement resolution (0.5 μm) of the LRSs, the displacement accuracy (1 μm) of the motion stage, the influence of temperature drift or ground vibration, and the installation repetitive error of the BPM. Among them, the first three are small enough so as to be ignored, while the BPM installation errors will affect the electrical center parameters X_{bpm_E} , Y_{bpm_E} , and the mechanical center parameters X_s , Y_s . which means that installation errors are the main source of random errors.

Installation errors resulting in imperfect parallel within the system that causes rotation errors in in the X, Y, and Z planes of the BPM coordinate system. Theoretically, if steps 3-4 in table 1 are strictly executed, all installation errors will be eliminated, but the calibration efficiency will be seriously reduced in actual engineering. Therefore, in order to improve calibration efficiency, the BPM installation errors were analyzed without performing steps 3-4, and the results are shown in table 2. The error generated by rotation around the Z-axis is shown in figure 6.

Hence, excluding the errors that can be software eliminated, the measured random errors can be expressed as Eq. (6):

$$\Delta Y^{r}_{\text{offset}} = \Delta X^{rx}_{bpm\_E} + \Delta X^{ry}_{bpm\_E} + \Delta X^{rz}_{bpm\_E} + \Delta X^{rz}_{s}$$
$$\Delta X^{r}_{\text{offset}} = \Delta Y^{rx}_{bpm\_E} + \Delta Y^{ry}_{bpm\_E} + \Delta Y^{rz}_{bpm\_E} + \Delta Y^{rz}_{s} \tag{6}$$

Assuming that $\theta_z = 0.1°$, $l_3 = 4$ mm, $l_4 = 50$ mm, and $H = 210$ mm, the error value $\Delta X^{rz}_{s} = 7$ μm and $\Delta Y^{rz}_{s} = 87$ μm will be obtained. Thus, due to the large l_4, slight rotation of the Z axis will cause a large ΔY^{rz}_{s}, which should be avoided as much as possible. In order to eliminate the installation rotation error, it is necessary to install BPM with a fixed torque wrench.

Table 1. The steps to determine the BPM mechanical center

Step Number	Operating Items	Operation Purposes
1	Move the LRSs in the Z-direction from the leftmost to the rightmost of the antenna support, and observe whether the data of the two LRSs change	Ensure that the movement direction of the LRSs is parallel to the two positioning planes (shown in figure 4) of the antenna support. Then, calculate X_{bpm_M} and Y_{bpm_M} based on Eq. (3).
2	Install BPM. Let the antenna through the BPM vacuum chamber and press it on the positioning planes of the antenna support.	Ensure that the antenna between the two pillars of the support is parallel to the two positioning planes.
3	Move the LRSs in the Z direction, scan the top and sides planes (shown in figure 4) of the BPM, and observe whether the readings of the two LRSs change.	Ensure that the axis of the BPM is parallel to the antenna (Eliminating the rotational errors around X-axis and Y-axis)
4	Hold the LRSs at the middle position of the antenna support, move BPM along the Y-axis, and observe whether the reading of the No. 2 LRS changes	Ensure that BPM is placed vertically (Eliminating the rotational error around Z-axis)
5	Adjust the position of BPM so that the distances between BPM and LRSs are equal to X_{bpm_M} and Y_{bpm_M}, which means that the antenna is in the BPM mechanical center at this time.	Align the antenna with the BPM mechanical center, open the automatic calibration software and start the calibration program.
6	Adjust BPM and let the antenna reach its electrical center, read the two LRSs data X_{bpm_E} and Y_{bpm_E}, and calculate offsets according to Eq. (4).	Only measure the electro-mechanical offset

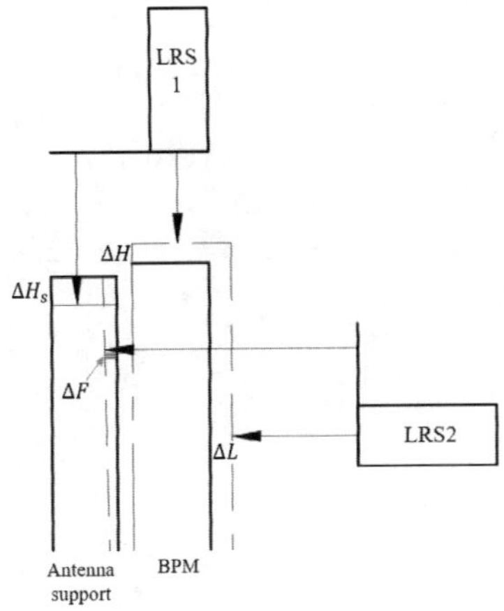

Figure 5. Schematic diagram of system error.

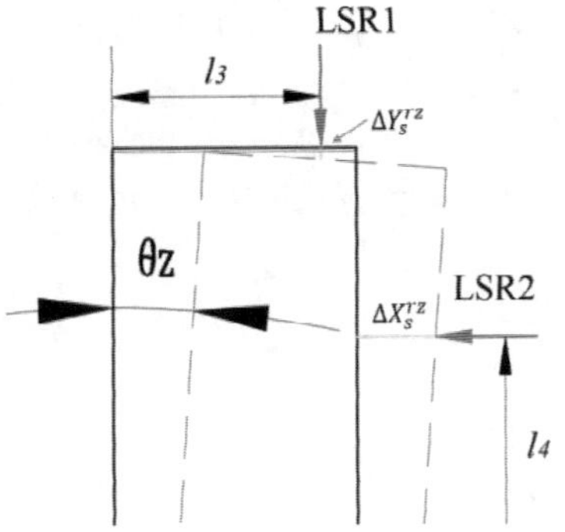

Figure 6. The error model for BPM rotation around the Z-axis

Journal of Physics: Conference Series **2681** (2024) 012003 doi:10.1088/1742-6596/2681/1/012003

Table 2. BPM installation errors analysis

Error Source	Effect on Measurement	Elimination Method
Rotation around X-axis	$\Delta X^{rx}_{bpm\_E}$ is difficult to evaluate. $\Delta Y^{rx}_{bpm\_E}$ is difficult to evaluate.	Hardware: Tighten the BPM mounting screws using a fixed torque wrench
	$\Delta Y^{rx}_{s} \approx l_1\theta_x$, l_1 is the distance from the laser spot to the X-Y plane. θ_x is the rotational error.	Software: The LRS1 scans along the Z-axis and the error ΔY^{rx}_{s} can be eliminated by averaging the measured Y_s at multiple positions.
Rotation around Y-axis	$\Delta X^{ry}_{bpm\_E}$ is difficult to evaluate. $\Delta Y^{ry}_{bpm\_E}$ is difficult to evaluate.	Hardware: Tighten the BPM mounting screws using a fixed torque wrench
	$\Delta X^{ry}_{s} \approx l_2\theta_y$, l_2 is the distance from the laser spot to the X-Y plane. θ_y is the rotational error.	Software: The LRS2 scans along the Z-axis and the error ΔX^{ry}_{s} can be eliminated by averaging the measured X_s at multiple positions.
Rotation around Z-axis	$\Delta X^{rz}_{bpm\_E}$ is difficult to evaluate. $\Delta Y^{rz}_{bpm\_E}$ is difficult to evaluate.	Hardware: Tighten the BPM mounting screws using a fixed torque wrench
	$\Delta Y^{rz}_{s} \approx l_3\theta_z - H\theta_z^2/2$, l_3 is the distance from the laser spot to the Y-Z plane. θ_z is the rotational error. $\Delta X^{rz}_{s} \approx l_4\theta_z$, l_4 is the distance from the laser spot to the X-Z plane.	No feasible software elimination method

4. Experiment and discussion

The physical experimental platform designed according to figure 3 is shown in figure 7. To improve the calibration efficiency, the two LRSs and precision motion stages are controlled by a computer program. The parameters of the tested BPM are $H = 210$ and $L = 12$. Support parameter $H_s = 182$ and the diameter of the antenna $r = 0.105$. By completing step 1 in table 1, we can get $X_s = 5.820$, $Y_s = 1.780$, $X_{bpm\_M} = -0.287$, and $Y_{bpm\_M} = -0.328$. The unit of all above data is mm. Then, install a BPM and adjust its position until the antenna is located at the electrical center, and measure X_{bpm_E} and Y_{bpm_E}. Repeat the experiment 5 times in three different ways and observe the repeatability accuracy. The experimental data is shown in table 3.

Table 3. The experimental data

Measurement operation	σ_{x_offset}(μm)	σ_{y_offset}(μm)
Repeat installation of the BPM without changing the status of the antenna	3.3	10.2
Repeat installation of the antenna without changing the BPM status	8.1	2.4
Repeat the installation of the antenna and BPM simultaneously	15.4	12.2

Figure 7. The experimental platform

Table 3 shows that changing only the antenna or BPM results in small random errors, and the repeatability accuracy is within 10 μm. If both of them are changed simultaneously, the repetition precision is approximately 15 μm. The more objects that are changed, the more uncertainties are introduced and the greater the measurement error is obtained. In practical engineering, each calibration requires manual operation of the BPM and antenna at the same time, so the third row in table 3 is the most realistic. On the basis of three-sigma rule, the range of measurement results is mostly within ±30 μm.

5. Conclusion

In order to quickly, low-cost, and high-precision measure the BPM electro-mechanical offset, a new method using precision antenna support and laser ranging sensor is proposed. The system's mechanical components are designed, and an automatic measurement program was developed. The resolution and maximum absolute error of this method are 15 μm and 20 μm, respectively. Improving the machining accuracy and surface finish of the antenna support, and designing a more reliable antenna pressing mechanism can further improve the calibration accuracy.

Acknowledgments

This work was funded by the Foundation of Youth Innovation Promotion Association, CAS (Y202005), and the Major achievements cultivation project of major scientific and technological infrastructure, CAS(NE01G74Y2), and the National Natural Science Foundation of China (11805221).

Journal of Physics: Conference Series **2681** (2024) 012003 doi:10.1088/1742-6596/2681/1/012003

References

[1] He J, Sui Y F, Li Y, Tang X H, Wang L, Liu F, Liu Z, Yu L D, Liu X Y, Xu T G, Cao J S and Yue J H. 2022 *Nucl. Sci. Tech.*, **33** 141.

[2] Chung Y, Decker G, Kahana E, Lenkszus F, Lumpkin A and Sellyey W. 1993 *Proc. Int. Conf. on Particle Accelerators (IPAC)*, Washington, DC, USA p 2304.

[3] Chen J Y, Chen C H, Chiu M S, Chou P J, Kuo C C, Liang C C, Liu Y C, Tsai H J and Tseng F H. 2019 *Nucl. Instrum. Methods Phys. Res., Sect A*, **932** pp 56-61.

[4] Tang X H, He J, Yue J H, Wei S J, Du Y Y, Ma H Z, Liu Z, Yang J, Gao G D, Cao J S, Wang A X, Sui Y F and Ye Q. 2022 *Nuclear Techniques*, **45**(2) 020102.

[5] He J, Sui Y F, Li Y, Wang A X, Tang X H, Zhao Y, Ye Q, Ma H Z, Xu T G, Yue J H and Cao J S. 2023 *Nucl. Instrum. Methods Phys. Res., Sect A*, **1045** 167635.

Canadian Light Source beam position visualization tool

M A Bree, T D Batten, J M Vogt and C W Stevens

Canadian Light Source, Saskatoon, Canada

E-mail: michael.bree@lightsource.ca

Abstract. The Canadian Light Source orbit correction system acquires, collates, and publishes storage ring beam centroid position information from 48 beam position monitors (BPMs) at a rate of 1000 samples per second. We present a "Beam Position Visualization Tool" that computes and displays dynamic Fast Fourier Transforms (FFTs) and Cumulative Power Spectral Densities (CPSDs) for all BPMs in real-time using full resolution data. The computed FFTs and CPSDs can be plotted in various combinations and in waterfall plots that allow visualization of changes over long periods of time. In addition, correlations between all BPM channel combinations are computed and ranked. Data from any two BPM channels can be selected for plotting in two dimensions wherein correlations are visually apparent. Computed CPSDs are further binned and archived for further analysis. Preliminary results from the Beam Position Visualization Tool have proven useful in identifying storage ring beam position noise at the Canadian Light Source.

1. Introduction

The Canadian Light Source (CLS) is a third generation synchrotron commissioned in 2005 that now supports 22 operational beamlines. The storage ring orbit correction (OC) system was upgraded in November 2021 to replace obsolete hardware and support new diagnostic capabilities [1]. The original Beam Position Monitors (BPMs) and associated Bergoz electronics are unchanged but new high speed D-tAcq analog-to-digital converters (ADCs) [2] feed beam position data via fibre optic cables to an OC host computer running Scientific Linux 7. The OC host uses the BPM data as input to the orbit correction algorithm to compute and send setpoints to the orbit corrector magnets (OCMs) over dedicated private Ethernet networks. The OC system uses the Experimental Physics and Industrial Control System (EPICS) [3] for configuration and control.

Many new features were built into the new OC system to facilitate specific diagnostic requirements such as hardware status. One such feature, called *beam position broadcast*, publishes a data stream consisting of the beam positions read at all OC BPMs. Since the data stream protocol is connectionless and static, its usage by clients is completely decoupled from the OC system. Future diagnostic tools can be implemented using the data stream without modifying, interrupting, or otherwise interacting with the running OC system in any way.

This paper covers two general topics:

- *Beam position broadcast*, which publishes BPM data.
- *Beam position visualization tool*, which is an example of a client that consumes the published BPM data.

2. Beam position broadcast

The new OC system uses four D-tAcq ACQ435ELF ADCs to read BPM values of the X (horizontal) and Y (vertical) beam centroid positions at 48 locations around the CLS storage ring. The BPMs are read at 1000 Hz, and each reading (sample) is three bytes. Using four bytes per sample results in a data rate of 384,000 bytes/s (plus protocol overhead) to transmit all 96 channels.

Publishing the beam position data in some form allows clients external to the OC system to process the information as they see fit, independently of OC operation. The CLS uses the EPICS control system and one possible mechanism for publishing beam position data is via EPICS Process Variables (PVs). EPICS clients can connect to PVs via a TCP-based protocol and can receive (and transmit) data via PVs over the control system network. However, the 1000 Hz update rate is high for EPICS (although there are techniques to mitigate this). More important, though, was the desire to not burden the real-time OC host computer that publishes the BPM data stream with hosting EPICS PVs for full resolution BPM data; and to not require data consumers (clients) to support EPICS. Therefore, EPICS was not used to publish the BPM data stream.

2.1. System architecture

The User Datagram Protocol (UDP)[4] was chosen to publish the BPM data stream. UDP is a lightweight, connectionless protocol that minimizes the burden on the OC host computer which has as its primary task running the orbit correction algorithm and sending updates to the OCMs. A dedicated CPU core on the OC host was assigned to transmit the data stream on a configurable network interface. The data stream is transmitted on a private network to keep it off the control system network. It is sent to the broadcast address which allows any number of clients to independently receive it with no impact on the transmitting OC host. For example, the beam visualization tool described later in this paper is a data stream client that runs on a separate computer which is connected to the private network. The tool is a pure Python program that receives the BPM data stream directly, processes it, and publishes its results as EPICS PVs. The overall system architecture is illustrated in Fig. 1.

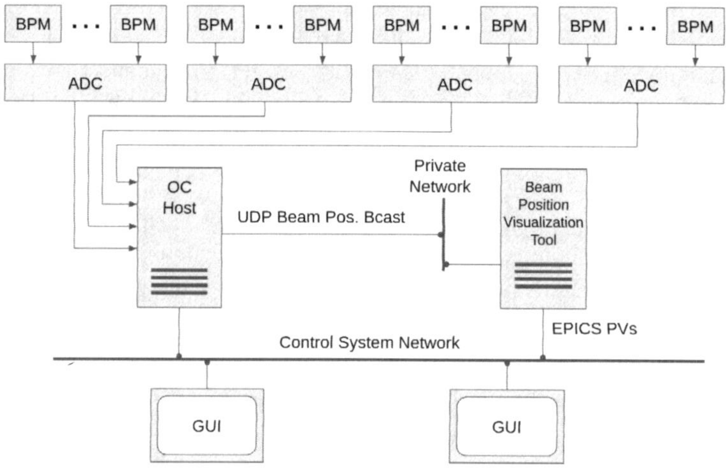

Figure 1. System architecture: Each of the 4 ADCs supports 12 BPMs. The OC host transmits the BPM data stream on a private network on which multiple clients can receive it. Processed results are published as EPICS PVs on the control system network where they may be accessed by screens and graphical user interfaces (GUIs).

2.2. BPM data stream protocol

Each transmitted frame consists of a 20-byte header as shown in Fig. 2. Typically, clients read the header of each received frame, and immediately reject any frame that does not contain the 32-bit identifier. Next they can read the 32-bit length field which indicates the number of bytes in the frame and reject any frame with a size that does not match. They may optionally check the 64-bit sequence number as frame delivery and frame order is not guaranteed by UDP[1]. For example, FFT computation could be paused if missing data is detected. Finally, the 32-bit command is read and the frame processed accordingly. Currently, two commands are implemented:

- **BPM_Position:** The frame contains sequential 32-bit numbers, each representing the calibrated voltage read by a BPM ADC.
- **BPM_Meta:** The frame contains a JavaScript Object Notation (JSON) string consisting of a human readable dictionary specifying the index, name, and volt-to-metre conversion factor for each channel. The dictionary also contains the current sampling rate. Such dictionaries are easily parsable by high level languages such as Python.

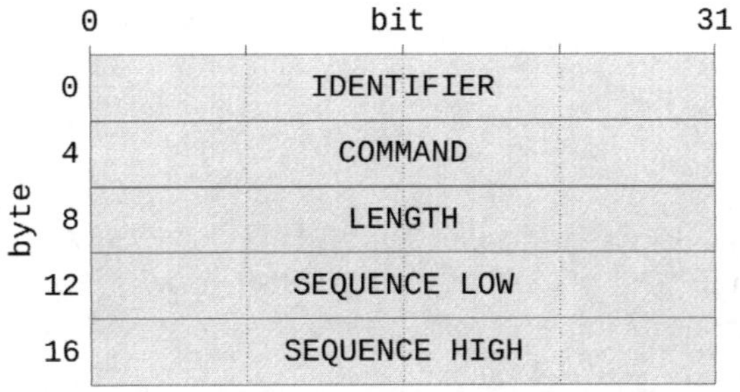

Figure 2. BPM data stream protocol frame header structure.

The OC host transmits a **BPM_Position** frame each time the BPMs are read (1000 Hz). Also, it sends a **BPM_Meta** whenever the meta data change, or once per second otherwise. This protocol simplifies client implementation. For example, clients need not be EPICS aware as they would be had EPICS been chosen to publish the BPM data.

3. Beam position visualization tool

3.1. Dynamic fast fourier transforms

After the new OC system was commissioned, a project was launched to use the new BPM data stream to dynamically compute and display the Fast Fourier Transforms (FFTs) and Cumulative Power Spectral Densities (CPSDs) of all 96 BPM channels. A pure Python multi-threaded program reads the data steam and waits for the next **BPM_Meta** frame that when received allows the program to dynamically configure itself. It then sends subsequent BPM samples from **BPM_Position** frames to worker threads, one for each channel. By monitoring the frame sequence numbers, missing data can be detected and FFT computations suspended until the data stream resumes.

The Python `multiprocessing` module is used for each worker, ensuring true parallel processing. Each worker adds each beam position sample to 10,000 element First-In-First-Out (FIFO) memory buffers for FFT computations. This produces a FFT from 0.1 to 500.0 Hz at 0.1 Hz resolution. At some configurable interval computations are performed using the FIFOs as input and the resulting FFTs are published as EPICS waveform PVs.

[1] In 1.5 years of continuous operation no missing or out of sequence frames have been detected.

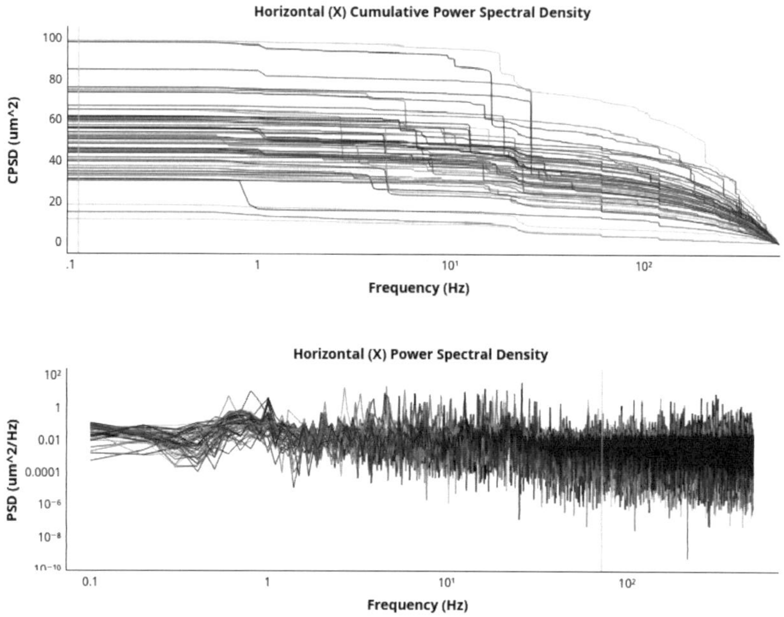

Figure 3. CPSDs and FFTs for all 48 horizontal BPM channels [1]. 60 Hz noise can be seen in the CPSDs for all channels. Other large spikes affect only individual BPMs or small groups of BPMs. Any combination of channels may be selected for plotting.

The FFTs are integrated from the Nyquist frequency (500 Hz) back to DC resulting in CPSDs which are also published as EPICS waveform PVs. The FFT and CPSD waveform PVs are large and there are 96 of each type. Therefore, the system implements a watchdog PV; any client (such as a GUI) must periodically write to the watchdog PV otherwise publication of the waveform PVs is suspended, minimizing control system network traffic. A GUI (Fig. 3 and first presented in Ref. [1]) was developed to display the dynamically computed FFTs and CPSDs. The dynamic display makes it easy to see transient noise, and noise that affects single BPMs and multiple BPMs. For example, in Fig. 3, 60 Hz noise is discernible in all BPMs.

3.2. Data archiving

The CPSDs are further summed into five large frequency bins: 0.1-1 Hz, 1-10 Hz, 10-100 Hz, and 100-500 Hz. Each of these values is published at a configurable interval (currently once per second) as an EPICS scalar PV. This results in 480 PVs (i.e., 5 per channel) suitable for archiving.

3.3. Dynamic waterfall plots

Another visualization technique is to treat each FFT for a BPM as a row in an image with pixel intensity corresponding to FFT amplitude. The resulting images illustrate noise as a function of time on the vertical axis and frequency on the horizontal axis. A new row is added to the top of the image each time a new FFT is calculated, and the bottom row (based on a configurable time span) is removed. Thus the image scrolls downward, leading to the *waterfall* name. Some interesting results are shown in Figs. 4 and 5. These images were collected with orbit correction in operation and with typical user-mode ring current. At the present time, sources of noise revealed by this new visualization tool are under investigation.

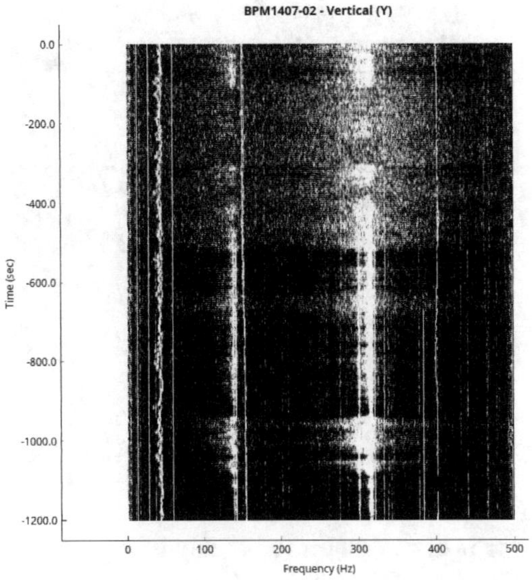

Figure 4. Intermittent noise across a wide frequency spectrum.

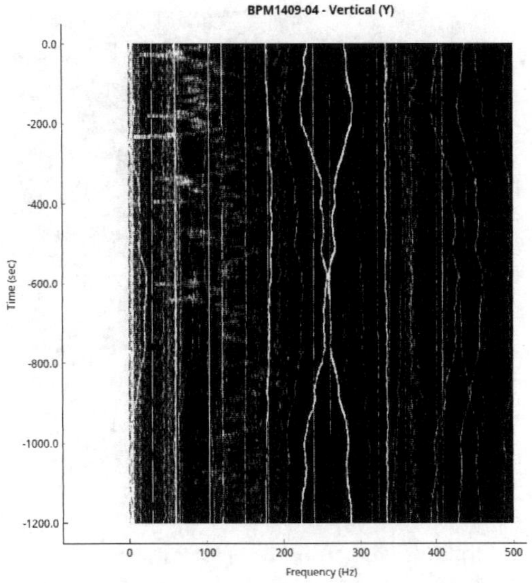

Figure 5. Narrowband noise drifting between 200 Hz and 300 Hz.

3.4. Position visualization
The tool uses the BPM data stream to plot data from pairs of BPM channels as pixels on an X-Y grid. The grid is divided into bins, and a bin "hit count" is incremented each time a measured position falls within it. As data is acquired over time, an image appears with pixel intensity proportional to the hit count, resulting in probability density function of beam position. An example is shown in Fig. 6.

Bin hits can be optionally aged out, and the pixel takes on the behaviour of a phosphorous screen with persistence. What appears is a dynamically changing image of beam position, updated at 1000 pixels per second. When the X and Y channels of the same BPM are chosen, the pixels represent the actual measured beam centroid position at the BPM. Examples are illustrated in Figs. 7 and 8.

3.5. Channel correlation
The tool computes the correlation between all combinations of the 96 beam position channels over sliding time windows of configurable duration. The pairs with the highest correlation are displayed on a GUI from which they can be selected for plotting.

3.6. Machine learning
The CLS is researching the potential application of machine learning techniques to orbit correction [5]. A Python client was written to gather and store full resolution beam position data from the BPM data stream for training the neural network. Users can now gather beam position data whenever and for whatever duration they want.

4. Future work
The data stream could be expanded to include full rate orbit corrector magnet (OCM) setpoints. This would include potential commands such as **OCM_Setpoint** and **OCM_Meta**. These are analogous to the **BPM_Position** and **BPM_Meta** commands described above except that they would describe the OCM setpoints computed by the OC system and the OCM metadata

Journal of Physics: Conference Series **2681** (2024) 012004 doi:10.1088/1742-6596/2681/1/012004

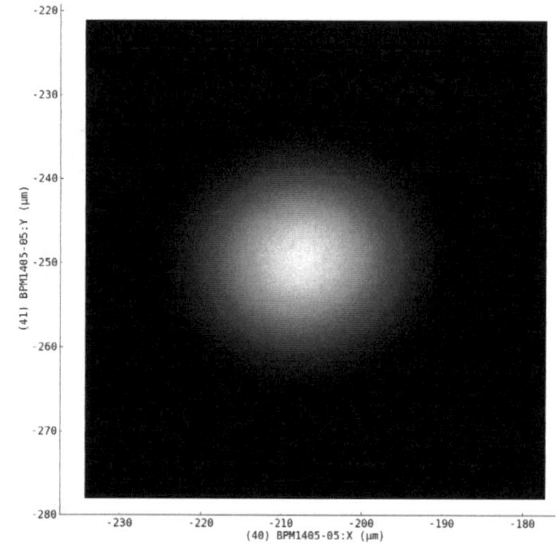

Figure 6. Image of beam position at BPM1405-05 with aging disabled. All recorded positions are counted and the pixel intensity reflects the total number of observations in the corresponding bin.

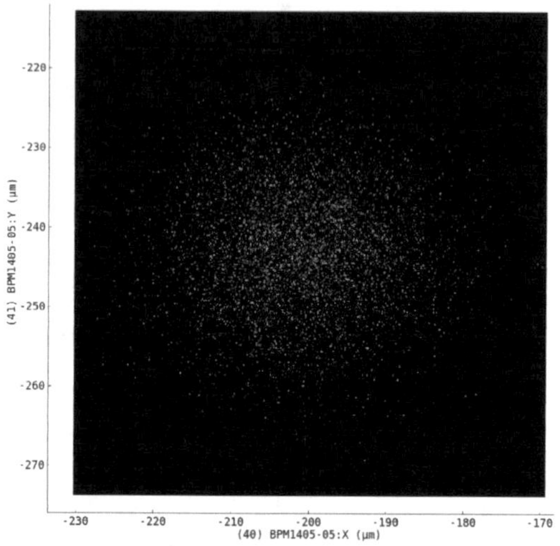

Figure 7. Image of beam position at BPM1405-05 with aging enabled. That is, position observations are discarded after a configurable period of time. The dynamic image is updated at 1000 pixels per second.

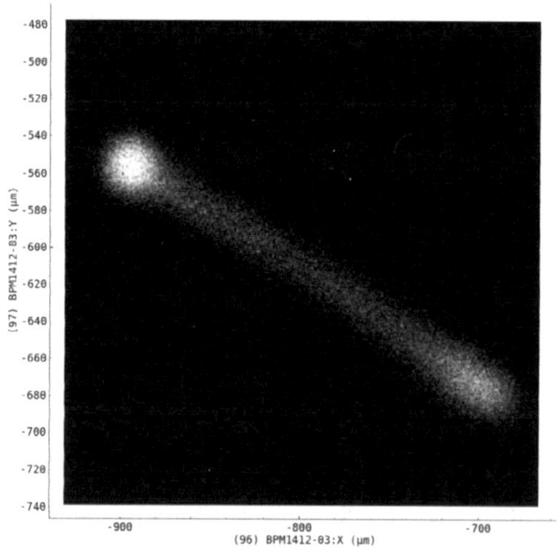

Figure 8. Beam motion at BPM1412-03 during injection with ring current at 3.4 mA.

respectively. This additional data, along with the beam position information, has the potential to facilitate external tools to characterize the performance of the complete OC system. There are many possibilities for analysis of the streamed data. For example unexpected beam behaviour such as clustering or deviation from other reference measurements could be detected and flagged. Finally, although not strictly part of this visualization effort, the sources of the noise revealed by the waterfall plots should be investigated.

5. Conclusion

The beam position broadcast feature has facilitated new visualization and analysis tools after the deployment of and without affecting the in-production orbit correction system. The extremely simple protocol allows anyone with interest to receive and analyze the BPM data stream (e.g., with simple Python) with no risk or impact on the orbit correction system.

The results have provided new insights and views of the CLS beam position characteristics that are still being evaluated. The archived CPSD values have proven useful for correlating changes in beam noise to other events, including modifications to the orbit correction algorithm. Likewise data gathered from the BPM data stream has facilitated studies such as machine learning initiatives.

Acknowledgments

Research at the CLS is funded by the Canada Foundation for Innovation, the Natural Sciences and Engineering Research Council of Canada, the National Research Council of Canada, the Canadian Institutes of Health Research, the Government of Saskatchewan, Western Economic Diversification Canada, and the University of Saskatchewan.

References

[1] Batten T D, Bree M A and Vogt J M 2022 Orbit correction upgrade at the Canadian Light Source *Proc. 11th Int. Beam Instrum. Conf. (IBIC'22)* (Kraków, Poland) pp 485-487 `doi:10.18429/JACoW-IBIC2022-WEP34`
[2] D-tAcq ACQ435ELF 32 channel simultaneous digitizer `https://www.d-tacq.com/acq435elf.shtml`
[3] Experimental Physics and Industrial Control System (EPICS) `https://epics.anl.gov`
[4] User Datagram Protocol (UDP) `https://www.ietf.org/rfc/rfc768.txt`
[5] Saadat S and Boland M J 2023 A new approach for CLS future orbit correction system driven by neural network *Proc. 12th Int. Beam Instrum. Conf. (IBIC'23)* (Saskatoon, Canada) `doi:10.18429/JACoW-IBIC2023-MOP036`

A new approach for Canadian Light Source future orbit correction system driven by neural network

S Saadat[1] and M J Boland[1,2]

[1] Canadian Light Source, Saskatoon, Canada
[2] University of Saskatchewan, Saskatoon, Canada

E-mail: shervin.saadat@lightsource.ca

Abstract. The orbit correction system (OCS) of the Canadian Light Source (CLS) comprises of 48 sets of BERGOZ beam position monitors (BPM). Each BPM has the ability to measure the position of the beam in both the X and Y directions and can record data at a rate of 1000 times per second. Inverse Response Matrix (IRM) is utilized to determine the optimal strength of the 48 sets of orbit correctors in both the X and Y directions, in order to ensure that the beam follows its desired path. The suggestion in this study is to replace the singular value decomposition (SVD) function with a neural network algorithm, which will act as the central processing unit of the orbit correction system. The training model's design includes three hidden layers, and within each layer, there are 96 nodes. The neural network's outputs for regular operations in CLS exhibit a mean square error (MSE) of 10^{-7}. Various difficult scenarios were created to test the OCS at 8.0 mA, using offsets in different sections of the storage ring. However, the new model was able to produce the necessary orbit correctors (OC) signals without any trouble.

1. Introduction

The CLS synchrotron light source storage ring consists of twelve sections and operates at a 2.9 GeV energy level. To maintain the stability of the beam position in the storage ring, an Orbit Correction System is utilized to correct any disturbances.

In 2000, the Motorola single-board computer was the initial implementation of a real-time controller [1]. It utilizes the Matlab channel access library to establish a connection with the fundamental EPICS control system [2]. CLS developed and tested a new orbit correction system the CLS Matlab application, known as CLSORB, with a high-speed capability [3]. This system offerd an adjustable rate range of 20 Hz to 100 Hz. The OCS at CLS comprised a computer that ran Matlab, along with four Versa Module Eurocards (VMEs). Each VME corresponded to three sections of the storage ring. Additionally, the system also incorporated a Real-Time Executive for Multiprocessor System (RTEMS). In 2021, the OCS underwent a significant upgrade [4]. As part of this upgrade, the previous RTEMS-VME system was retired and replaced with new components. The recently developed orbit correction system integrates D-tAcq Analog to Digital Converters (ADCs) for monitoring BPM positions. It also features Nuclear Instrument Modules (NIM) utilizing PoLabs single board controllers for the parallel distribution of magnet power supply setpoints and a scientific Linux 7 operating system is used to run the new software. The purpose of this advanced system is to ensure that any disturbances caused by electron perturbations are immediately detected and corrected to maintain beam stability and optimal light quality at the beamlines. Hardware and software of the orbit control systems were developed until the RMS deviation of beam motion was reduced to less than one micrometer in both the X and Y directions. The Accelerator Operations and Development (AOD) team is currently working on upgrading this system. This involves improving its efficiency, increasing its accuracy and adding new features to meet the evolving needs of users. With technological advancements being made every day, it is important to keep upgrading existing systems to stay relevant and efficient in a rapidly changing landscape. The AOD team has embarked on new research to design a Dynamic Orbit Correction System (DOCS) based on Neural Network (NN) algorithm for the orbit correction system. The NN correction system offers advantages over the IRM algorithm, excelling in dynamic adaptability, flexible programming, and computation speed. Unlike the static IRM, the NN system learns from data, making it accurate in changing conditions. Its adaptable nature handles misalignments effectively. The NN's efficient script allows customization and rapid computations, while IRM's matrix recalculations slow it down. The precision and real-time capabilities of this DOCS have been reported [5]. The aim of this research is to improve the accuracy of the neural network model by incorporating deep learning techniques. To achieve this goal, we will conduct experiments and analyze the data to determine the effects of deep learning methods on accuracy.

2. Neural network

A neural network is a type of machine-learning model inspired by the structure and function of the human brain. It consists of interconnected nodes or neurons that process information to make predictions or decisions [6]. Also, Deep learning is a subfield of machine learning that uses neural networks with multiple layers to extract and learn features from data. It is capable of automatically discovering complex patterns and relationships in large datasets [7].

The following sections will briefly discuss the neural network architecture and parameters utilized in the CLS orbit correction system model, and also provide a definition of deep learning loops.

2.1. Network architecture

Designing a neural network architecture requires a significant amount of information about the real system it will be applied to, as the architecture is highly dependent on the specific characteristics and structure of the system. In brief, the CLS orbit correction system is comprised of 96 Bergoz Beam Position Monitors (BPMs) that transmit the beam's position to the IRM at a frequency of 1000 measurements per second. The IRM serves as the central processing unit of the OCS, responsible for computing the strength of the Orbit Correctors. Ultimately, the 96 outputs from the IRM are transmitted to the 96 orbit correctors at a rate of 18 transmissions per second.

In the first step, we designed a neural network algorithm instead of using IRM. For this case, we utilized the TensorFlow-Keras library in Python, which is renowned for its speed and is considered one of the fastest AI modules. The neural network consists of three hidden layers with 96 nodes, and the best activation function was found to be hyperbolic tangent. After a single learning step with 20 epochs, the model achieved a MSE of $(3.152 * 10^{-7})$. The Mean Squared Error is a loss function that measures the average squared difference between the predicted values and the actual target values in the training dataset.

2.2. Deep learning loops

The initial model performed well, and its real-time capabilities were tested on actual data, exhibiting negligible error when the system was in a stable mode. In addition to the aforementioned scenario, there is another situation that should be taken into account, when the beam position changes significantly. In this case, the neural network model must be capable of adapting to these changes in order to provide accurate predictions. Therefore, it is essential to ensure that the model is trained using a diverse set of data that encompasses a wide range of beam positions and conditions. This will enable the model to effectively learn and generalize to new and unseen scenarios, including cases where the beam position changes drastically. In this study, a learning loop was designed to address the issue at hand. The loop begins by loading the initial model, after which a new dataset is evaluated using the model. If the error of the model's output exceeds the expected value, the model is updated with new epochs and

Journal of Physics: Conference Series **2681** (2024) 012005 doi:10.1088/1742-6596/2681/1/012005

activation functions as necessary. Finally, Fig. 1 the accuracy of the updated model is checked, and it replaces the previous model if it performs better.

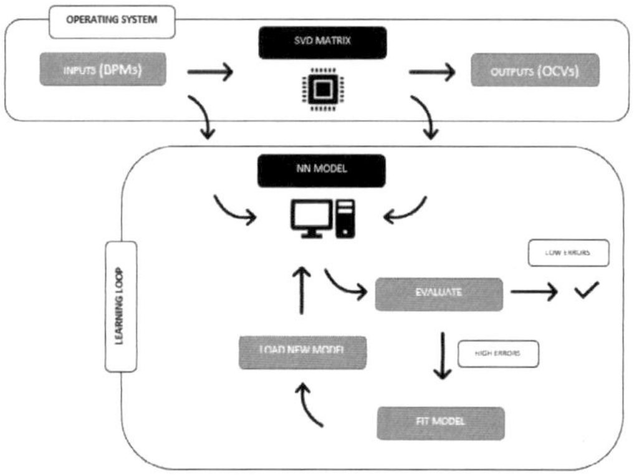

Figure 1. The BPM and OC datasets from the real system are fed into the learning loops. The NN model uses the BPMs to generate new OCs, and the error between the predicted and real OCs is calculated. If the error exceeds the expected threshold, the NN model is updated with the current dataset and replaces the old NN model. This loop is repeated periodically to refine the NN model.

3. Experiments and results

The CLS storage ring is comprised of twelve distinct sections, each of which contains four BPMs and four OCs with both x and y directions. This configuration yields a total of 96 inputs and 96 outputs for the orbit correction system. Figure 2 shows a schematic of BPMs configuration in the storage ring. Ninety-six inputs, which correspond to BPMs, are recorded at a rate of 1000 Hz, while the system generates ninety-six outputs, which correspond to OCs, at a rate of 18 Hz. There are two scenarios that we have assumed for the system. In the first scenario, the system operates in a stable mode, and the learning loop attempts to improve its model accuracy. In the second scenario, the stable mode changes to varying positions, and the learning loop modifies the

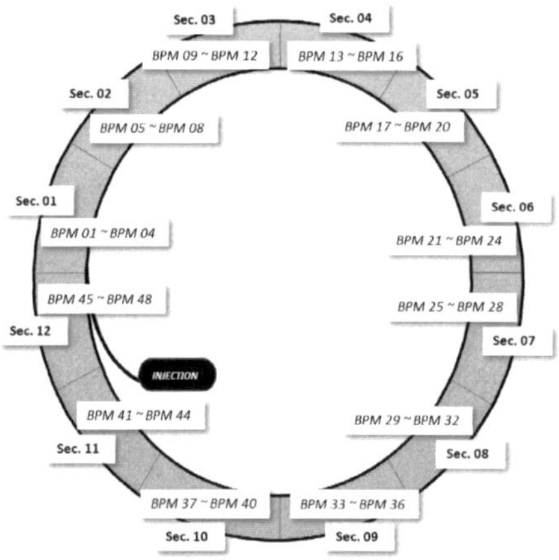

Figure 2. A schematic of the CLS (Canadian Light Source) storage ring that comprises twelve sections and 48 BPMs.

model to prepare for the future. In this research, we investigated the deep learning capabilities of the DOCS platform. Due to the lack of available real-time data, we collected data from the CLS using a sampling frequency of 1000 Hz for BPMs and 18 Hz for OCs. The collected data was saved in large JSON files, which were then split into smaller files for use in our deep-learning model.

In the first scenario, when the system is functioning normally, we conducted the learning loop four times. During the first phase, we developed the model, while the second to fourth phases involved updating the model. In addition, we introduced a metric called Accumulated Error (AE), which represents the sum of errors across all OCs. The calculation of AE involves taking the absolute difference between the real value of the orbit corrector and the predicted OC, divided by the real data. Our developed model demonstrates excellent performance, with an AE of 109.72%. When applied to the 96 OCs, the error rate per OC is estimated to be approximately 1.1%. Figure 3(a) suggests that the model is robust and reliable in predicting values for the given dataset. We implemented the deep learning loops step by step, beginning with an upgrade without any changes in the epochs and architecture. Remarkably, the AE decreased to 73.06%, indicating a significant improvement in performance. This corresponds to an error rate of approximately 0.76% per orbit corrector, which is substantially lower than the previous error rate. Figure 3(b) demonstrates the efficacy of the deep learning approach in improving the accuracy of the model. Continuing with the upgrade, we observed in the Fig. 3(c) a further improvement in the model's performance. The AE decreased to 50.59%, which equals an error rate of approximately 0.53% per OC. 4th upgrade, we observed in the Fig. 3(d) again more improvement in the performance. The AE decreased to 40.48%, or 0.42% per OC.

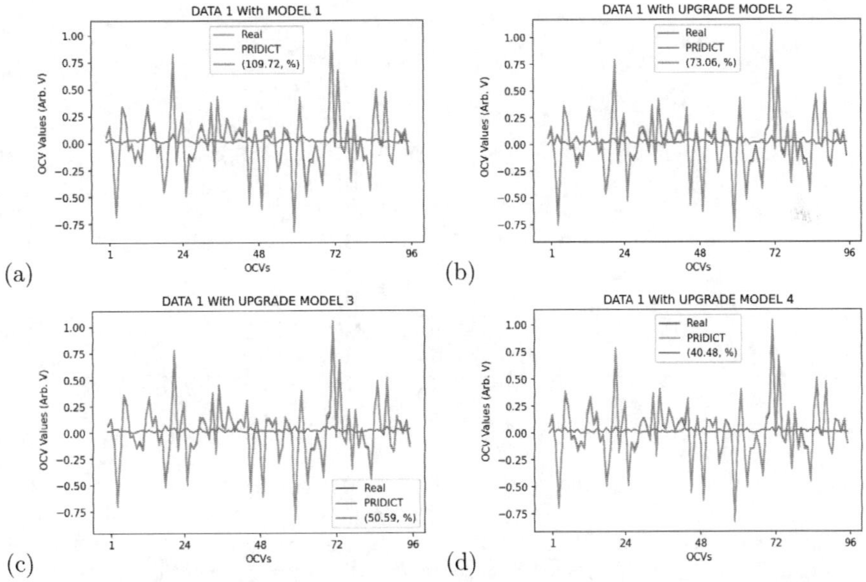

Figure 3. The blue line represents the actual data for 96 OCs obtained through the SVD algorithm. The orange line represents the predicted values for the same OCs obtained through NN models. The green line shows the difference between the actual and predicted values. The green value corresponds to the total error (AE) calculated across all 96 OCs. The horizontal axis shows the 96 orbit correctors. (a) The first model (b) The first upgrade (c) The 2nd modification (d) The 3rd upgrade.

In the following scenario, the beam position is expected to undergo a sudden change, presenting a challenging situation for the OCS. We aim to investigate whether this loop can update a model with memorized coefficients, furthermore, we will calculate the error that arises during this process. The objective of this experiment is to evaluate the effectiveness of the

Journal of Physics: Conference Series **2681** (2024) 012005 doi:10.1088/1742-6596/2681/1/012005

updates to the NN model, and to assess whether they result in a decreasing or increasing trend in performance.

At the CLS, a total of four BPMs and four OCs are installed in each section of the storage ring. Each BPM measurement includes both Horizontal (H) and Vertical (V) beam position. Specifically, BPM01 - 04 are located in section one, and BPM05 - 08 are installed in section two, and similarly, the BPMs configuration is repeated throughout the remaining sections as shown in Fig. 2.

In this experiment, the BPM's values in different sections of the storage ring (SR) were changed by adjusting the reference (offset) to create a desired beam position situation (+1 and -1 mm) for the OC system. The system was tested at a beam current of 8.0 mA. Table (1) displays the alterations to the reference (offset) and how they reveal variations in BPM values across distinct SR sections.

Table 1. Some of the adjustments made to the reference or offset (μm) were shown to demonstrate the change in BPM values across different sections of the SR. 'H' and 'V' refer to the horizontal and vertical detectors, respectively. Rand refers to random offsets utilized for all BPMs, ranging between -1000 μm and 1000 μm.

Experiment No.	Section	BPMs	Offset μm
1	1	H BPM 01	+1000
		V BPM 01	-1000
2	1	H BPM 01	+500
		V BPM 01	+500
		H BPM 01-08	+500
		V BPM 01-08	+500
		H BPM 09-16	-1000
		V BPM 09-16	-1000
		H BPM 17-24	+500
19	ALL	V BPM 17-24	+500
		H BPM 25-32	-1000
		V BPM 25-32	-1000
		H BPM 33-40	-1000
		V BPM 33-40	-1000
20	ALL	H BPM 01-48	Rand.
		V BPM 01-48	Rand.

The OCS's performance was recorded in the aforementioned scenarios for both BPMs and OCs. This data was subsequently utilized to learn, train, test, and develop DOCS.

In this research, we tested several assumptions and found that the results of all of them were the same. Here, we present one of these results. In the first scenario, the beam is in the position that corresponds to test 1 in the table (1). The NN model was developed (with 80 epochs and the same architecture) for this situation, and we consider it to be our initial model which has 1.184% error per OC as shown in Fig. 4(a). The learning loop attempts one upgrade and generates an NN-upgraded model with 0.515% error per OC as shown in Fig. 4(b).

Journal of Physics: Conference Series **2681** (2024) 012005 doi:10.1088/1742-6596/2681/1/012005

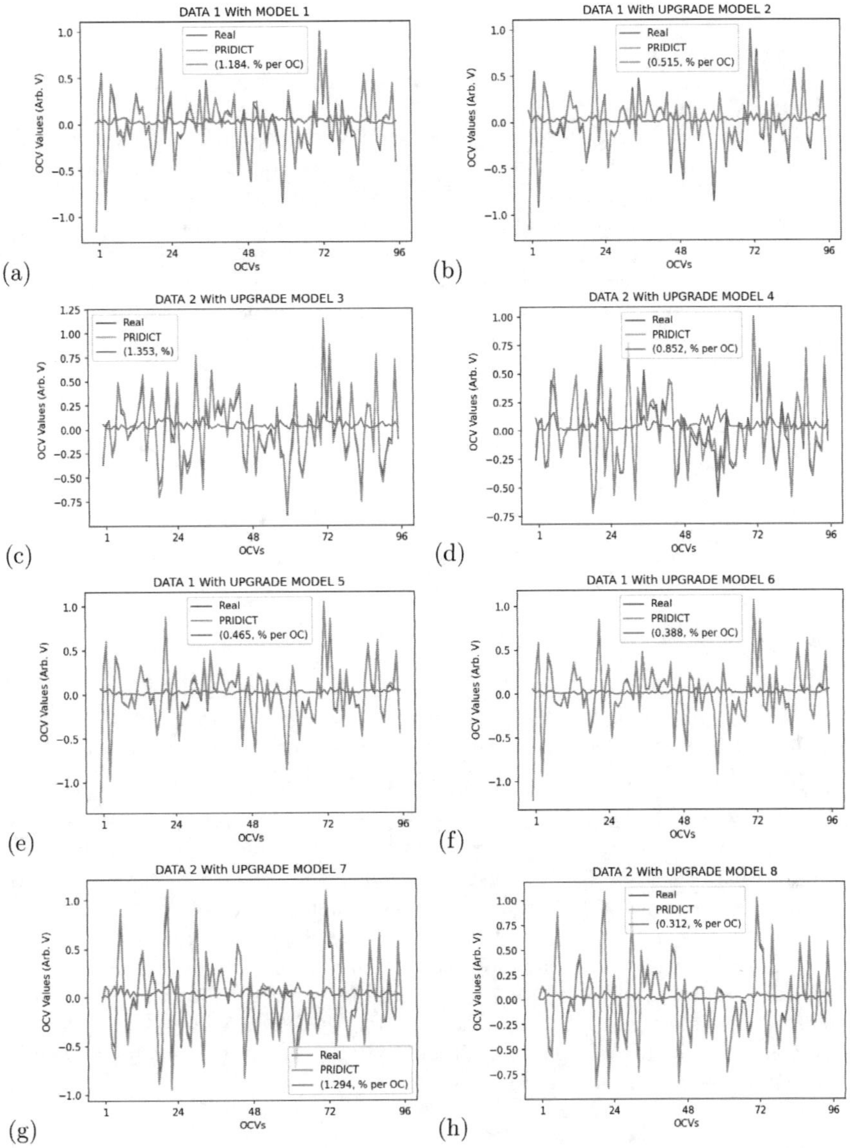

Figure 4. The blue line represents the actual data for 96 OCs obtained through the SVD algorithm. The orange line represents the predicted values for the same OCs obtained through NN models. The green line shows the difference between the actual and predicted values. The green value corresponds to the mean error for each OC. The horizontal axis shows the 96 orbit correctors. (a) The first model (b) 1st upgrade (c) 2nd upgrade (d) 3rd upgrade (e) 4th upgrade (f) 5th upgrade (g) 6th upgrade (h) 7th upgrade.

Following that, a disturbance occurs, and the system transitions into the mode detailed in test 20 of Table 1. The NN model was updated twice during this period, as depicted in Figs. 4(c) and 4(d). The turbulence is controlled, and the perturbation sequence returns to mode 1, the NN model has upgraded twice again shown in Figs. 4(e) and 4(f). Finally, the system experiences a period of instability, during which the perturbation sequence exhibits unpredictable behavior. As a result, the learning loop generates two new updates for the NN model, as shown in Figs. 4(g) and 4(h). The scenario described, demonstrates that updates to the NN model's weight parameters are effective, resulting in a statistically significant decrease in error rates over time.

4. Conclusion
Two conditions of the NN model's effectiveness were investigated, a stable mode and a complex mode. The results for the stable mode demonstrate that error rates decrease over time, from 1.1% per OC in the first experiment to 0.42% per OC after four steps. While this result was expected, it provides further confirmation of the effective operation of the deep learning loops in the NN model, which successfully reduce error rates over time in the stable mode condition. The results of the second scenario are less definitive than those of the stable mode upgrade, as shown in Fig. 5. However, they do display a decreasing trend in error rates despite some fluctuations.

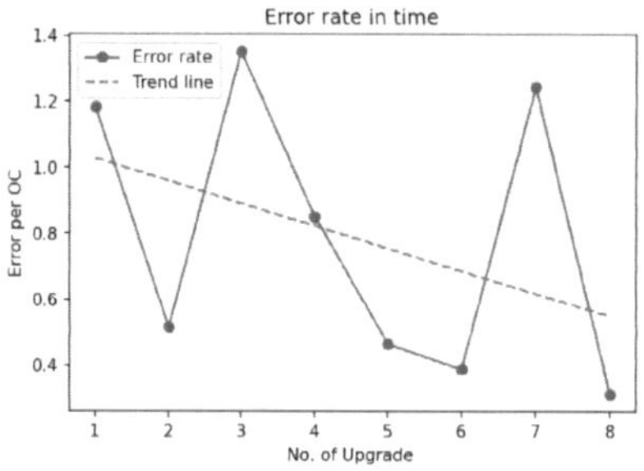

Figure 5. Errors (percentage per OC) are displayed in eight upgraded steps. The blue dots represent the mean errors in each step, the red line shows the trend of the errors.

Figure 5 shows a decreasing trend in prediction errors for a specific data set (upgrades 1 and 2), but the deep learning model is unable to generalize its performance to new data sets (upgrade 3) without further training. However, next step training result in reduced errors on new data sets (upgrades 4 and 5). This cycle continues, and the errors decrease smoothly over time. The results demonstrate that the NN model is highly effective, and the Deep Learning loop successfully increasing the model's precision. While the results show promise and suggest that the NN model could be a valuable technology for various applications, it is important to acknowledge the limitations of our study and the need for further research to thoroughly explore its capabilities and potential applications. The current findings, while encouraging, should be viewed with a degree of restraint, and additional studies are warranted to establish the full extent of the NN model's benefits and limitations.

Acknowledgments
The authors would like to thank Tonia Batten, Johannes Vogt, Ward Wurtz, Cameron Baribeau, Hamed Shaker, Drew Bertwistle, Mike Bree, and Fredric Le Pimpec, as well as all other AOD members of the CLS operation group.

References
[1] Dallin L, Blomqvist, I, de Jong M, Lowe D, and Silzer M 2003 The Canadian Light Source *Proc. PAC'03* (Portland: OR, USA) TOPA001
[2] Chabot D 2008 SR1 Orbit Control Design Note *CLS Document* 5.2.39.4
[3] Payne C G and Chabot 2010 Fast Orbit Correction at the Canadian Light Source *Proc. PCaPAC'10* (Saskatoon: Canada) WECOMA02 p 9
[4] Batten T, Bree M and Vogt J M 2022 Orbit Correction Upgrade at the Canadian Light Source *IBIC'22* (Krakow: Poland) WEP34 p 485
[5] Saadat S and Boland M J 2023 Model of a dynamic orbit correction system based on neural network in CLS *IPAC'23* (Venice: Italy) WEPA094 p 2862
[6] Hagan M T, Demuth H B, Beale M H and De Jesus O 2002 *Neural network design* (Hagan.Okstate Edu.)
[7] Goodfellow I, Bengio Y and Courville A 2016 *Deep learning* (MIT Press)

Journal of Physics: Conference Series **2681** (2024) 012006 doi:10.1088/1742-6596/2681/1/012006

Coupled bunch mode zero correction within the orbit feedback bandwidth

P Kallakuri, A Brill, J Carwardine, L Emery and N Sereno
Advanced Photon Source, Argonne National Lab, Lemont, USA
E-mail: pkallakuri@anl.gov

Abstract. The fast orbit feedback (FOFB) bandwidth for Advanced photon source upgrade (APS-U) will be DC-1 kHz and the synchrotron frequency will be between 100-560 Hz. This frequency overlap places coupled bunch mode 0 (CBM0) induced horizontal orbit motion inside the orbit feedback bandwidth, potentially affecting our ability to achieve beam stability goals. Longitudinal feedback kicker is not strong enough to damp CBM0 oscillations. We developed new beam-based feedback method to suppress CBM0 oscillations with low level RF phase as actuator. It uses existent FOFB framework with no changes to the feedback algorithm. Effectiveness of this method is verified using present APS operations lattice where synchrotron frequency is outside orbit feedback bandwidth. In the present work, low alpha lattice is created to emulate APS-U setting where synchrotron frequency is inside the orbit feedback bandwidth. Experiments with this lattice successfully demonstrated CBM0 correction within orbit feedback bandwidth. Combined operation of orbit feedback and CBM0 correction is stable, and CBM0 oscillations are damped. We achieved better orbit motion suppression and corrector drive efforts are reduced as well.

1. Introduction

The target bandwidth of the Advanced Photon Source Upgrade (APS-U) Fast Orbit Feedback (FOFB) is DC-1000 Hz where the synchrotron tune will be between 100-560 Hz. This overlap places Coupled Bunch Mode 0 (CBM0) induced horizontal position offsets within the FOFB bandwidth range, affecting APS-U goals for beam stability. Large storage-rings such as APS would need longitudinal feedback system with high kick voltage capability for CBM0 suppression. APS-U longitudinal feedback kicker is not strong enough to damp CBM0 oscillations. We developed new orbit to RF phase feedback method using existent orbit feedback framework with no additional processing hardware requirements [1]. It operates at the Low Level RF (LLRF) signal level of the main RF system and does not require high kick voltages. Based on synchrotron oscillation theory [2, 3], we derived relationship between beam position deviation at dispersive BPMs and RF phase noise that represents the open loop dynamics of our feedback configuration. An analytical model is devised for the coupling mechanism of synchrotron oscillations to transverse orbit through dispersion. It allows an energy-induced component to be extracted from the measured orbit which is derivative of the RF phase error. Using this the orbit feedback controller generates a RF phase control signal as another drive signal in the orbit feedback algorithm. Experiments conducted at the present APS storage ring using 7 GeV operations lattice and APS-U prototype fast orbit feedback controller [4, 5] demonstrated the effectiveness of this method at damping coupled bunch mode zero oscillations.

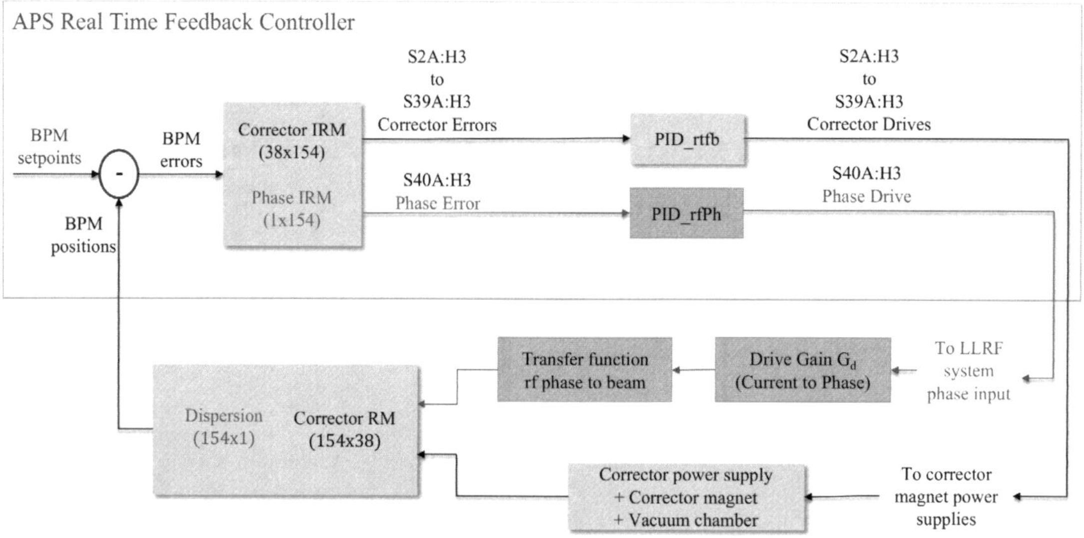

Figure 1. Experimental setup for simultaneous operation of RTFB and orbit to RF phase feedback using APS RTFB system.

In the above experiments [1] the synchrotron frequency (2.2 kHz) and orbit feedback bandwidth (920 Hz) are well separated. We continued our study using low-alpha lattice configuration developed to mimic APS-U setting where, synchrotron frequency (60 Hz) is within the orbit feedback bandwidth (90 Hz). We would be able to study the interaction of both feedbacks more clearly in this setup. The storage-ring energy is set to 6 GeV and low momentum compaction factor of $3.6 \cdot 10^{-06}$ is used to get synchrotron frequency to around 60 Hz. Orbit feedback with 1.5 kHz sampling rate used for APS operations is termed as Real Time Feedback (RTFB) [6]. Experiments with low-alpha lattice are conducted using RTFB system so that we would be able to use 38 fast correctors and 154 BPMs around the ring. With the prototype system we could only use 3 fast correctors and 12 BPMs in 2 sectors that is not adequate to deal with larger noise seen in low-alpha lattice. Details of our experimental setup with RTFB controller, closed loop performance of orbit to RF phase feedback, and results from the simultaneous operation of RTFB and CBM0 correction with low alpha lattice are presented in next sections.

2. Experimental Setup

Closed loop schematic of RTFB together with orbit to RF phase feedback is shown in Figure 1. RTFB controller hardware only allows use of total 160 BPMs (4 per sector) for orbit feedback computations. We used 154 BPMs for beam position measurements. Orbit feedback uses 38 horizontal fast correctors, and orbit to RF phase feedback is integrated into RTFB framework by repurposing fast corrector (S40A:H3) analog drive signal as phase drive. DAC output channel of S40A:H3 is connected to LLRF system [7] phase input. The cavity phase loop bandwidth is adequately greater than the synchrotron oscillation frequency. So the RF system is considered as simple gain term in the feedback model. PID_rtfb controller is used orbit correction and PID_rfPh controller is used for CBM0 correction. Same feedback algorithm that generates corrector power supply set-points is used for phase drive. Phase computations are incorporated as additional row in Inverse Response Matrix (IRM) dot product. Horizontal BPM position vector $\mathbf{x}$ is obtained from 154 BPMs. Phase error is calculated as dot product of measured orbit and weighted dispersion vector *Phase_IRM* as defined in Eq. (1). α_c is the momentum compaction factor, ω_{rf} is RF angular frequency, and η is the dispersion vector.

$$Phase\_IRM = \frac{180}{\pi}\frac{1}{G_d}\frac{\omega_{rf}\alpha_c}{\eta \cdot \eta}\eta \tag{1}$$

Drive gain $G_d = \frac{1}{13}$ is to convert phase drive in amperes to RF phase set-point in degrees. Phase IRM (1x154) computed from Eq. (1) and Corrector IRM (38x154) obtained from orbit response matrix are concatenated as 39x154 IRM used in the experiments. For beam-based studies we used 6 GeV low alpha lattice with storage-ring current of 14 mA in 24 bunch operation and 24 mA in 48 bunch operation. Relatively low currents are used since it was very difficult to inject into this lattice. So, beam loading is negligible.

3. Orbit to RF phase feedback for CBM0 correction

First step in our experiments is to establish LLRF phase set-point control from RTFB controller in open loop, and then verify the stability of orbit to RF phase feedback and suppression of synchrotron frequency magnitude in closed loop. For open loop measurements S40A:H3 drive

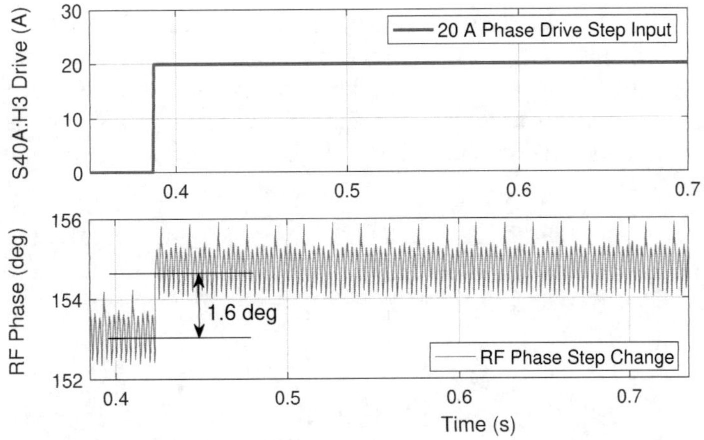

Figure 2. Open loop phase step response measured for 20 A input step amplitude.

signal is the input, and beam phase response is the output. Figure 2 shows the step input to phase drive, and measured beam phase response. Input step of 20 A resulted in 1.6° change in RF phase response confirming drive gain $G_d = \frac{1}{13}$. Large 360 Hz noise from klystron power supply is seen in the steady state RF phase.

3.1. CBM0 Correction in Closed Loop Operation

After testing the open loop drive setup, studies are conducted with closed loop orbit to RF phase feedback configuration using 1x154 phase IRM. The storage ring configuration for this test is 24 bunches, 14 mA, 6 GeV low alpha lattice with synchrotron frequency at 60 Hz. Closed loop is stable with negative proportional gain and we achieved suppression around synchrotron frequency. We tested K_p gains in the range -0.01 to -0.06, the beam was unstable beyond -0.06. Motion at 360 Hz.

Figure 3 shows square root integrated Power Spectral Density (PSD) of BPM error for open loop and closed loop with different K_p gain values. Orbit motion suppression at synchrotron frequency (60 Hz) is observed when the feedback is ON. Increase in proportional gain resulted in more suppression at synchrotron frequency.

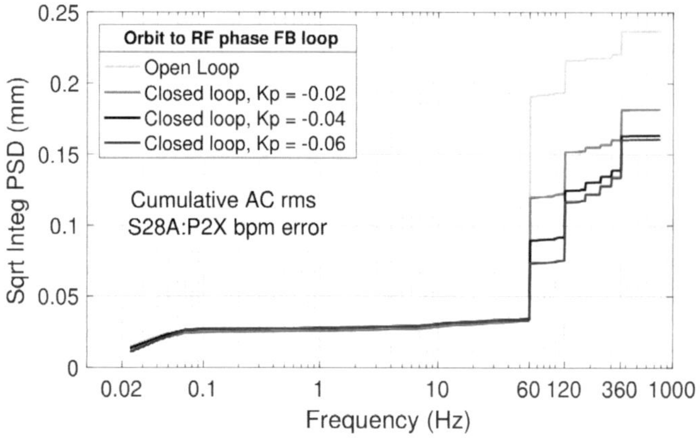

Figure 3. Orbit motion measured in open loop, and closed loop with different proportional gain values.

4. Real Time Feedback with Coupled Bunch Mode zero correction

Next step in our experiments is to study interaction between CBM0 correction and RTFB during the simultaneous operation of both feedbacks.The test configuration is 6 GeV low alpha lattice with 48 bunches and 24 mA storage-ring current. Synchrotron frequency 60 Hz is within the orbit feedback bandwidth of 90 Hz emulating APS-U settings.

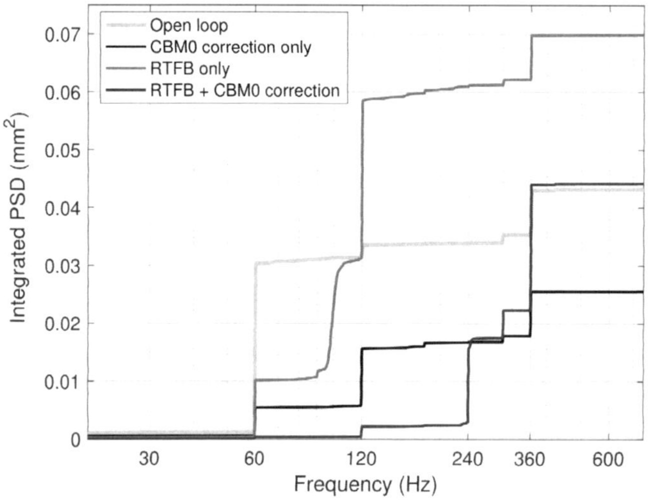

Figure 4. Comparison of cumulative mean square orbit motion at S28A:P2X BPM for different feedback configurations.

Closed loop with unified 39x154 IRM is stable with proportional controllers. For RTFB we used K_p gains in the range 0.01 to 0.03, and for orbit to RF phase feedback we used K_p gains in the range -0.01 to -0.05. For individual operation optimum performance is achieved with maximum permissible K_p gains in each case. For simultaneous operation we tuned the gains together for optimum performance. For RTFB $K_p = 0.02$, and for CBM0 correction $K_p = -0.03$ are the optimal gains. The angle between dispersion and betatron closed orbits with 154 BPMs are in the range $83-97°$ which is comparable to 440 BPMs case shown in Ref. [1]. We are working on a simulation study to further understand whether this near orthogonality is inferring complete decoupling when both corrections has a frequency overlap. BPM noise response data is measured for different feedback configurations: open loop, CBM0 correction only $K_p = -0.05$, RTFB only $K_p = 0.03$, RTFB $K_p = 0.02$ + CBM0 correction $K_p = -0.03$. Integrated PSD of BPM error in different feedback configurations is shown in Figure 4. Open loop orbit motion shows significant spike at 60 Hz, which is suppressed in all three closed loop configurations: RTFB only, CBM0 correction only, and RTFB + CBM0 correction. During individual operation, orbit to RF phase feedback and RTFB are correcting respective energy and betatron components at synchrotron frequency which resulted in partial suppression at 60 Hz. High frequency motion beyond 90 Hz is amplified during RTFB only operation. Motion at 120 Hz is amplified when only CBM0 correction is used. With both feedbacks running together we got significant suppression at 60 Hz due to the combined effect. Also, high frequency motion amplified by individual feedback operation is attenuated up to $\approx$220 Hz. Motion at 360 Hz is not suppressed by orbit to RF phase feedback. Perturbation at 360 Hz is in phase with RF phase modulation whereas perturbation at synchrotron tune is 90° out of phase. So, we would need large controller gains to affect 360 Hz which could be beyond the stability limit of synchrotron frequency damping.

BPM responses are measured at all 154 BPMs around the storage-ring to analyze envelope of the orbit motion. Figure 5 shows power spectral density plots of measured BPM errors in open loop and different closed loop configurations. All BPM responses showed similar frequency response pattern, while BPMs in the high dispersion area has larger magnitudes compared to others. Individual operation of RTFB or CBM0 correction partially suppressed 60 Hz and frequencies beyond 90 Hz are amplified. During combined operation significant suppression at 60 Hz and better orbit motion suppression up to 240 Hz is achieved. APS RF system has noise at 360 Hz and other 60 Hz harmonics. These frequencies have larger amplitudes in low alpha lattice configuration. Hence narrow resonant increase of noise at 60 Hz harmonics is seen when high frequency motion is amplified during individual and combined feedback operation.

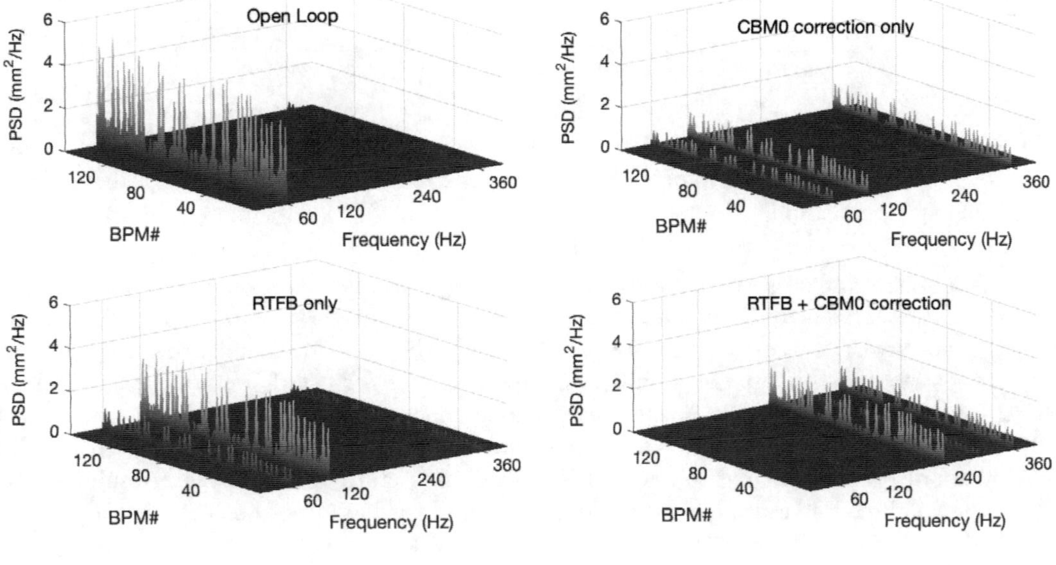

Figure 5. PSD of BPM errors at all 154 BPMs around the storage ring. Comparison of responses in open loop and different closed loop configurations.

Journal of Physics: Conference Series **2681** (2024) 012006 doi:10.1088/1742-6596/2681/1/012006

4.1. Analysis of Feedback Control Efforts

Feedback control effort is the input to the actuator from the dynamic controller to perform necessary correction. Reduction in control efforts is beneficial to the feedback operation.The corrector and phase drive signals in our case indicate the control efforts. In order to analyze the drive efforts of both feedbacks we used cumulative mean square of phase and 38 corrector drive signals. Integrated PSDs of drive signals during individual operation of each feedback and combined operation of both feedbacks are shown in Figure 6. Drive magnitudes during

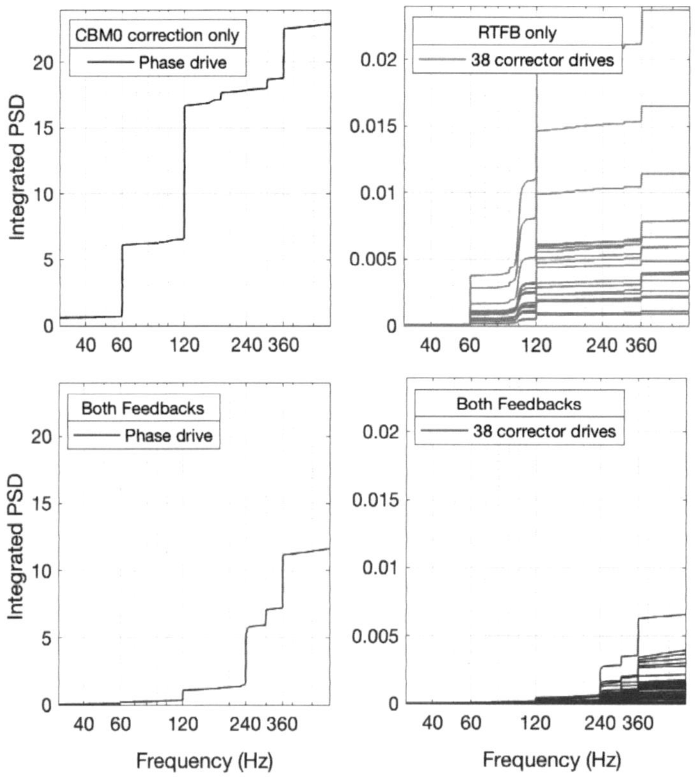

Figure 6. Comparison of drive efforts of each feedback individual operation with drive efforts in simultaneous operation.

individual operation are larger compared to simultaneous operation. During individual operation the feedbacks see large error at synchrotron frequency which results in more drive effort as the feedback keeps trying to correct the error. When CBM0 correction and orbit feedback are operated together both energy and betatron components are corrected simultaneously, and the error will be small. As a result phase and corrector drive magnitudes for frequencies below 120 Hz are significantly reduced in combined operation, and drive magnitudes up to Nyquist frequency are also less compared to respective individual operations. From these results we can say that the combined operation is aiding in control effort reduction of both feedbacks.

5. Conclusions

We successfully demonstrated CBM0 correction within the orbit feedback bandwidth. We used 6 GeV low-alpha lattice that provides a test setup where synchrotron frequency (60 Hz) is within the RTFB bandwidth (90 Hz) similar to APS-U configuration. First, we established LLRF phase set-point control from RTFB controller open loop and then verified the stability in closed loop. Orbit to RF phase feedback loop is stable and we achieved suppression around synchrotron frequency. Closed loop operation with both feedbacks using unified corrector and phase IRM is stable with proportional controllers. During individual operation, orbit to RF phase feedback and RTFB are correcting respective energy and betatron components at synchrotron frequency which resulted in partial suppression at 60 Hz. Orbit feedback was effective at reducing motion at synchrotron frequency, but did so at the expense of increasing broadband rms motion beyond the open-loop level. This was consistent with previous RTFB operating experience. Running CBM0 correction on its own also reduced motion at synchrotron frequency and without increasing the broadband rms motion. The combination of CBM0 correction and RTFB was significantly more effective than RTFB alone over all frequency bands from DC to 360 Hz. Feedback drive efforts are also reduced during combined operation of RTFB and CBM0 correction.

Acknowledgements
This work is supported by the U.S. Department of Energy, Office of Science, under Contract No. DE-ACO2-O6CH11357.
We are grateful to U.Wienands for the valuable insights. We wish to thank T. Berenc, S. Xu, S. Shoaf, R. Blake, H. Bui, and T. Madden for helping with experiments; V. Sajaev for the help with low alpha lattice setup.

References
[1] Kallakuri P, Brill A, Carwardine J and Sereno N 2022 *Phys. Rev. Accel. Beams* **25** 082801 doi:10.1103/PhysRevAccelBeams.25.082801
[2] Ormond W and Rogers T 1997 *Proc. Particle Accelerator Conference (Vancouver, Canada)* p 1822 https://jacow.org/pac97/papers/pdf/4V038.PDF.
[3] Wiedemann H 2015 *Particle Accelerator Physics 4th ed* (Springer) ISBN: 978-3-319-30759-6
[4] Sereno N et al 2015 *Proc. International Particle Accelerator Conference (Virginia, USA)* p 1167 doi:10.18429/JACoW-IPAC2015-MOPWI011
[5] Kallakuri P, Brill A, Carwardine J and Sereno N 2019 *Proc. North American Particle Accelerator Conference (Michigan, USA)* p 683 doi:10.18429/JACoW-NAPAC2019-WEPLM11
[6] Carwardine J and Lenkszus F 1998 *Proc. Beam Instrumentation Workshop (California, USA)*
[7] Song J.J et al 1998 *Proc. Asian Particle Accelerator Conference (Tsukuba, Japan)* p 767 https://accelconf.web.cern.ch/a98/APAC98/6D036.PDF

Journal of Physics: Conference Series **2681** (2024) 012007 doi:10.1088/1742-6596/2681/1/012007

Beam test of a harmonic kicker cavity

M W Bruker[1], J Grames[1], J Guo[1], J Musson[1], S A Overstreet[1], M Pablo[2], G T Park[1], T Plawski[1], M Poelker[1], R Rimmer[1], B Roberts[2], D Speirs[2], H Wang[1], C Wilson[1], S Zhang[1]

[1] Thomas Jefferson National Accelerator Facility, Newport News, VA, USA
[2] Electrodynamic, Albuquerque, NM, USA

E-mail: bruker@jlab.org

Abstract. A harmonically resonant kicker cavity designed for beam exchange in a circulator cooler was built and successfully tested at the Upgraded Injector Test Facility (UITF) at Jefferson Lab. This type of cavity is being considered for the injection scheme of the Rapid Cycling Synchrotron at the Electron-Ion Collider, where the spacing of neighboring bunches demands very short kicks. Operating with five transversely deflecting modes simultaneously that resonate at 86.6 MHz and consecutive odd harmonics thereof, the prototype cavity selectively deflects 1 of 11 electron bunches while leaving the others unperturbed. An RF driver was developed to synthesize phase- and amplitude-controlled harmonic signals and combine them to drive the cavity while also separating the modes from a field-probe antenna for RF feedback and dynamic tuning. Beam deflection was measured by sweeping the cavity phase; the deflection waveform agrees with expectations, having sub-nanosecond rise and fall times. No emittance increase is observed. Harmonically resonant cavities like the one described provide a new capability for injection and extraction at circulators and rings.

1. Introduction

In circulating accelerators, the minimum bunch spacing needed to accommodate the rise/fall time of a kicker can limit the design options for the bunch train. Originally developed for the Circulating Cooler Ring, a hypothetical ring with 11 revolutions driven by an energy-recovery linac that was intended to be part of the Jefferson Lab Electron-Ion Collider [1], the harmonic kicker cavity offers a new option for applications where every nth bunch must be deflected [2]. One such application is the injection into the Rapid Cycling Synchrotron at the Electron-Ion Collider [3], where one out of four consecutive bunches is injected at a time; here, the bunch spacing is 1.6 ns, and each set of four bunches is followed by a gap of 12 µs, encouraging the use of a pulsed device but out of reach for a stripline due to the short bunch spacing.

A harmonic kicker is a transversely deflecting cavity that can be excited at multiple harmonic frequencies at the same time, allowing one to Fourier-synthesize any deflection waveform containing only these frequencies; the high Q of the modes compared to a stripline makes the drive power manageable, while pulsed drive is still possible if the gap between sets of bunches is sufficiently long. Figure 1 shows a model of the prototype reduced to the most important parts.

An example waveform providing a kick to 1 out of 11 bunches is shown in Fig. 2. The usefulness of this type of waveform is not characterized by its absolute flatness outside of the main peak but by the locations of its zero crossings or minor peaks, depending on design. While conceptually flexible, the mode structure of such cavities must be chosen according to the bunch timing of the intended application. Compared to placing multiple cavities with different

Journal of Physics: Conference Series **2681** (2024) 012007 doi:10.1088/1742-6596/2681/1/012007

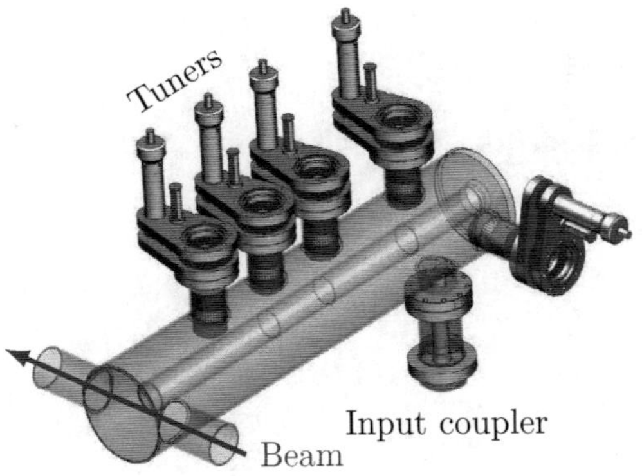

Figure 1. CAD model of a 5-mode harmonic kicker cavity. Five stub tuners are needed to tune all modes. The RF signal is coupled in through a single port; another port serves as the field probe (not shown here).

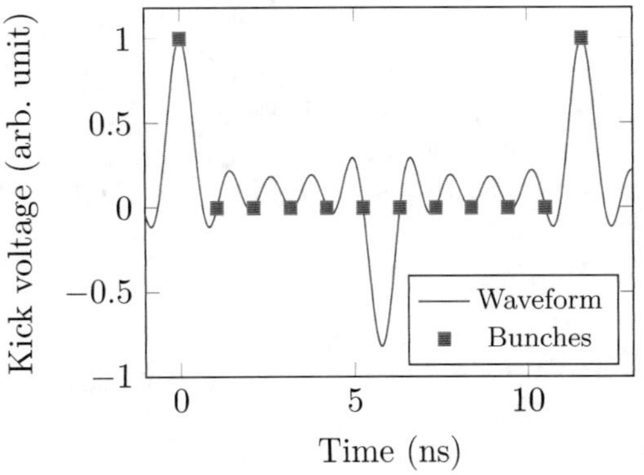

Figure 2. Kick action only on bunches at a bunch frequency equal to the fundamental of the kick waveform, $f_{HK} = 86.6$ MHz; in this example, all 11 buckets are filled at a bunch frequency of $11 f_{HK} = 952.6$ MHz.

resonant frequencies in series, combining the modes in one cavity has the upside of providing all the deflection in the same location, thereby avoiding bunch displacement due to the extra drift between cavities; however, note that only odd harmonics can readily be produced in a single cavity of this type.

Drawing from a diverse collaboration for fabrication [4], tuning [5], and drive-circuit design, we produced working prototypes of the cavity and the drive circuit and performed a beam test to verify the basic aspects of the cavity performance. The following sections describe the methods and results.

2. Harmonic Arbitrary Waveform Generator

The cavity is controlled by a specially designed drive circuit called the Harmonic Arbitrary Waveform Generator (HAWG). A synthesized RF source phase-locked to the master oscillator of the accelerator is used to generate an 86.6 MHz harmonic comb. The desired harmonic signals are divided from the harmonic comb onto individual channels using a harmonic combiner and divider [6]. Once isolated, each harmonic signal is phase- and amplitude-controlled using I/Q modulator ICs, amplified, and recombined to the cavity's drive antenna using a second harmonic combiner. While the cavity is designed for an input power of multiple kW, the present version of the HAWG provides 100 W per channel. With this power, the cavity produces its nominal deflection of ~ 4 mrad at 6.5 MeV/c.

Signals from the coupling antenna of the cavity are directed to a third harmonic divider where the harmonic signals are separated to individual I/Q demodulator ICs, allowing for both RF feedback and stub-tuner control for each mode.

3. Beam generation and synchronization

The beam test of the cavity was performed at Jefferson Lab's Upgraded Injector Test Facility (UITF). Designed to be accessible and reconfigurable with relatively little logistical effort, this accelerator delivers short (<1 ps) electron bunches at up to 8 MeV with an average current of 100 nA (limited only by radiation shielding). The beam is generated by a laser-driven DC photogun at 200 keV, ballistically bunched, and accelerated to the final energy by a compact, superconducting booster [7]. All RF elements run at $f_0 = 1497$ MHz; the signal pulsing the laser determines the bunch timing. In principle, the machine supports bunches at any subharmonic of f_0; also, configurable low-frequency (50–200 Hz) macropulses gating the drive laser can be set up regardless of bunch timing.

The frequencies useful for this experiment are determined by the cavity design. The cavity fundamental $f_{HK} \approx 86.6$ MHz is related to the machine base frequency f_0 by

$$\frac{f_{HK}}{f_0} = \frac{7}{121}. \tag{1}$$

The highest frequency that is a subharmonic of both f_{HK} and f_0 is $f_0/121 \approx 12.4$ MHz, making this frequency ideal for placing all bunches on the same cavity phase, whereas its 11th harmonic, $f_0/11 \approx 136.1$ MHz, could be used to probe all 11 phases of interest at the same time. The simplest way to measure the deflecting action of the kicker cavity is to send bunches at $f_0/121$; this way, all bunches are deflected the same, avoiding the complication of temporally selective detection downstream. The angle of deflection can then be measured as a function of relative phase by shifting the cavity waveform in time.

The frequencies needed for all RF systems (normally including the laser) come from the main RF synchronization chassis of the machine, which contains multiple independent phase-locked loops (PLLs) operating from a common 10 MHz reference. Because the $f_0/121$ signal needed for this experiment was not readily available from the laser PLL, we instead used a commercial frequency generator (CG635 by Stanford Research) synchronized to the same 10 MHz reference. Due to the nature of the frequency ratio, there is no way to produce the exact frequency with this generator, but with 16 digits of frequency resolution, the phase drift due to frequency mismatch is $< 0.1°$ per day.

Another CG635 set to the same frequency supplied the reference signal for the PLLs in the HAWG that generate all frequencies for the harmonic cavity, allowing for independent phase control of the HAWG with a single knob.

Apart from phase-synchronous RF drive of all components, an additional complication was the necessity for pulsed operation of the HAWG. While the device is designed for CW operation in principle, the duty cycle was limited to 10 % for this experiment as a precaution. In order to prevent bunches from being delivered with the cavity off, we operated the laser in macropulse mode with a pulse width of 400 µs and a delay of 100 µs with respect to the 50–200 Hz trigger signal, and gated the HAWG with a stretched copy of this trigger signal so that it would turn on sooner than the beam and turn off later, giving the cavity time to ring in. The timing is shown schematically in Fig. 3.

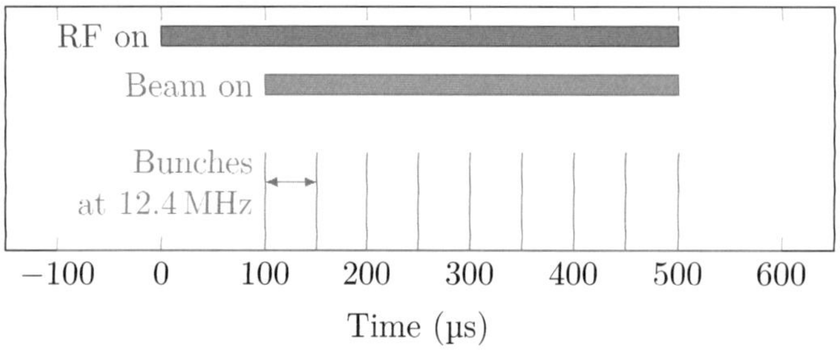

Figure 3. Macropulse timing for clean beam delivery while the cavity is on. Bunch spacing exaggerated for visual clarity.

The macropulse beam mode increases the bunch charge for the same average current, improving the signal-to-background ratio of the BPMs, which is too low for them to be sensitive in CW mode because of the average-current limitation of the UITF. On the other hand, because the macropulses are generated electro-optically with a finite contrast ratio of ≈ 250, very low macropulse duty cycles such as those needed for viewscreen-safe operation result in a high relative level of CW background during the nominally-off time window, obscuring the desired image on viewscreens after the kicker. This issue is not significant for other diagnostics because they can be used at higher beam duty cycles; also, the BPM data acquisition is gated to the macropulse signal and therefore insensitive to CW background.

4. Beam-line setup

For this experiment, we chose a beam momentum of 6.5 MeV/c for deflection angles $\leq$ 4 mrad at full HAWG output power. The deflection was detected by a pair of BPMs located downstream of the cavity. With the cavity off, the beam orbit was set up to be as flat as possible throughout this section and geometrically centered in the cavity. A pair of quadrupole magnets combined with a wire scanner located at the downstream end of the beam line was used to measure the transverse phase space; because the bunch frequency was chosen to place all bunches at the same kicker phase, the deflection from the kicker could be steered out with a pair of correctors, restoring the orbit through the emittance-measurement section. Figure 4 shows a schematic layout, while Fig. 5 shows a photo of the cavity installed in the UITF beam line.

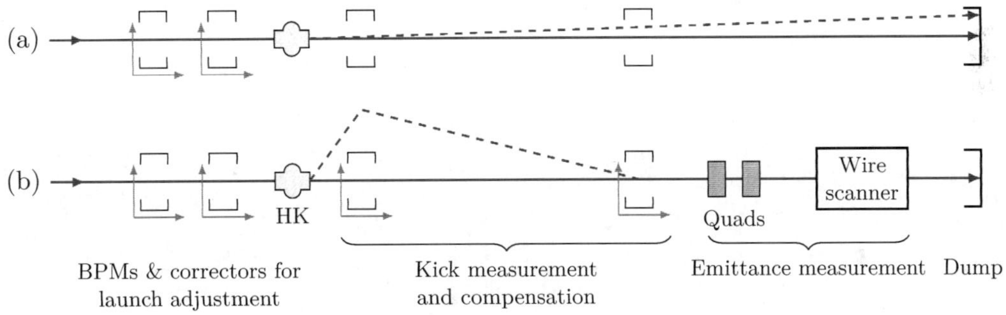

Figure 4. Conceptual sketch of the beam line for the kicker test. Part (a) shows the essential components of the setup for measuring only the deflection, while part (b) includes a kick correction with two steerers to restore the orbit for consistent quad scans. The dashed line represents the beam orbit with the kicker turned on; it is exaggerated for clarity in the bottom figure.

Figure 5. Harmonic kicker installed in the UITF beam line with the HAWG next to it.

5. Results

5.1. Deflection waveform

Due to time constraints, the beam test was performed without automatic stub-tuner control and RF feedback. However, after iterative matrix-based tuning of all modes and with temperature-regulated water flowing through the inner conductor of the cavity, we found that the frequency stability of all modes was well within the bandwidth. The limited RF duty cycle helped keep detuning due to RF heating to a minimum.

As a first test, we drove the cavity at half power per channel one mode at a time, the other amplifiers off, and scanned the HAWG reference phase to measure the deflection waveform at each frequency. The angle of deflection for each phase follows from the position signals of two subsequent BPMs located downstream of the cavity. Figure 6 shows an example of this measurement performed at the ninth harmonic, demonstrating the single-mode deflection is a clean sine wave.

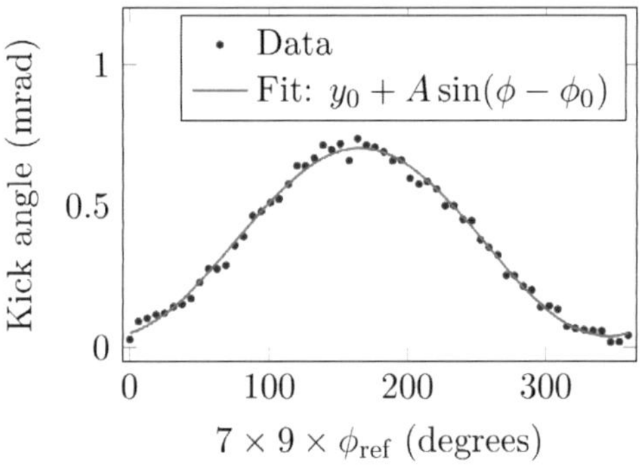

Figure 6. Deflection waveform with only the 9th harmonic (779.4 MHz) being driven at nominally 50 % output power. The phase ϕ_{ref} being scanned is that of the 12.4 MHz reference and is scaled up accordingly for clarity. The beam angle is not explicitly corrected to give zero with the cavity off. $y_0 = 0.372(3)$ mrad, $\phi_0 = 75.1(7)^\circ$, $A = 0.333(4)$ mrad.

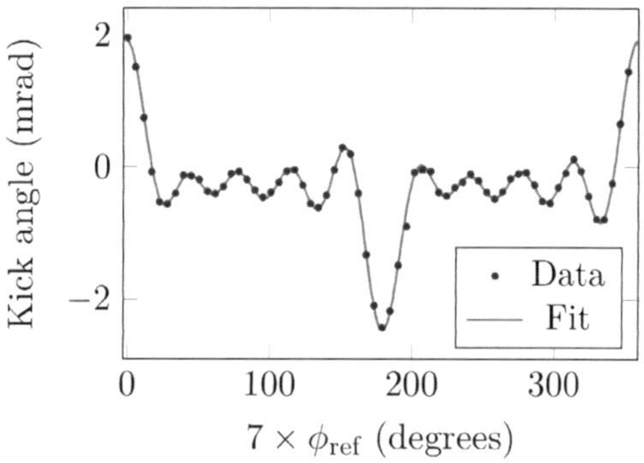

Figure 7. Deflection waveform with all five modes powered and optimized.

Having tested each mode individually, we powered all modes at the same time, measured the superposition waveform, and adjusted the phase and amplitude setpoints of the modes to give equal phases and amplitudes in the deflection waveform as determined by a fit of

$$f(\phi) = f_0 + \sum_{k=1}^{5} A_k \cos\left[(2k-1)\phi - \phi_k\right] \tag{2}$$

with $\phi = 7\phi_{\mathrm{ref}}$. With all A_k equal and all $\phi_k = 0$, this would give the desired waveform shown in Fig. 2.

The waveform so obtained is shown in Fig. 7, its parameters listed in Table 1. While the waveform visually resembles the target, the numbers show a discrepancy in the highest mode. The dominant limitation preventing an accurate and reproducible synthesis of the target waveform is the lack of both tuning feedback and RF feedback in the present setup, causing the phases to wander. In practice, this is exacerbated by each phase being slightly interdependent

Table 1. Fit parameters for waveform in Fig. 7

k	A_k (mrad)	ϕ_k
1	0.47	1°
2	0.46	2°
3	0.46	4°
4	0.43	4°
5	0.38	−16°

with the corresponding amplitude due to imperfections of the RF elements in the HAWG, especially in the highest mode.

The adjustability of the modes turned out to be further limited by multipacting, which could be observed as a vacuum response and intermittent jumps in the cavity field for several different combinations of RF parameters. While our attempts to process out the multipacting between beam shifts were unsuccessful, it is not expected to be present at higher levels of drive power as needed for the beam momenta of practical applications.

5.2. Transverse phase space

As is the case for all deflecting cavities, the time-dependent deflecting field of the harmonic kicker will transform the spread of arrival times—i.e., the bunch length—into a spread of transverse angles, deteriorating the 2-d emittance if the bunch length is significant compared to the temporal slope of the field around the central bunch phase. Multipole components of the field will further affect the transverse phase space if the transverse beam size is significant. As a sanity check to verify that neither effect is visible with the default bunch length of the UITF and an optics setup that ensures $\sigma_{x,y} \ll 1\,\text{mm}$, we performed quad scans at the end of the line at different cavity phases, correcting for the deflection with steering coils. The results are listed in Table 2, showing no measurable impact on transverse phase space regardless of kicker phase; the point at 0° is at peak deflection, whereas that at 15.4° is at maximum negative slope.

Table 2. Downstream Twiss parameters in the deflecting plane of the kicker

$7\phi_{\text{ref}}$	α	β (m)	ϵ (nm)
off	-9.9	21.4	26.3
0°	-9.6	19.9	26.7
15.4°	-9.5	20.0	24.4

6. Conclusion

We successfully operated the harmonic kicker with all five modes and performed a beam-based measurement and optimization of the deflection waveform, finding no unexpected issues other than multipacting. We expect the technical limitations of the drive circuit to be fully resolved by ongoing development efforts, making harmonic kicker cavities a viable option for future accelerator designs.

Acknowledgments

Work supported by the U.S. Department of Energy, Office of Science, Office of Nuclear Physics under contract DE-AC05-06OR23177. Multi-Harmonic driver development has been supported by DOE's NP SBIR program (DE-SC0020566).

References

[1] Benson S V *et al.* 2018 *Proc. COOL'17 (Bonn, Germany)* (Geneva: JACoW) pp 72–76 DOI:10.18429/JACoW-COOL2017-WEM12

[2] Park G T, Guo J, Rimmer R A, Wang H and Wang S 2021 *Phys. Rev. Accel. Beams* **24** 061002 DOI:10.1103/PhysRevAccelBeams.24.061002

[3] Willeke F and Beebe-Wang J 2021 *Electron Ion Collider Conceptual Design Report 2021* Brookhaven National Lab Report DOI:10.2172/1765663

[4] Overstreet S A, Bruker M W, Grose G A, Guo J, Henry J, Park G T, Rimmer R A, Wang H and Williams R S 2022 *Proc. NaPAC'22 (Albuquerque, NM, USA)* (Geneva: JACoW) pp 601–603 DOI:10.18429/JACoW-NAPAC2022-WEZD6

[5] Huang Y, Wang H, Rimmer R A, Wang S and Guo J 2016 *Phys. Rev. Accel. Beams* **19** 122001 DOI:10.1103/PhysRevAccelBeams.19.122001

[6] Roberts B F 2022 US patent 11 303 005: Harmonic combiner and divider

[7] Bruker M W *et al.* 2022 *Proc. NaPAC'22 (Albuquerque, NM, USA)* (Geneva: JACoW) pp 640–643 DOI:10.18429/JACoW-NAPAC2022-WEPA12

Design and study of cavity quadrupole moment and energy spread monitor

Q Wang[1], Q Dong[1], L Huang[1] and Q Luo[2]

[1] Dalian Institute of Chemical Physics, Chinese Academy of Sciences, Dalian 116023, China
[2] NSRL, University of Science and Technology of China, Hefei 230029, China

E-mail: luoqing@ustc.edu.cn

Abstract. A nondestructive method to measure beam energy spread using the quadrupole modes within a microwave cavity is proposed. Compared with a button beam position monitor (BBPM) or a stripline beam position monitor (SBPM), the cavity monitor is a narrow band pickup and therefore has better signal-to-noise ratio (SNR) and resolution. In this study, a rectangular cavity monitor is designed. TM220 mode operating at 4.76 GHz in the cavity reflects the quadrupole moment of the beam. The cavity plans to be installed behind a bending magnet in Dalian Coherent Light Source (DCLS), an extreme ultraviolet FEL facility. In this position, the beam has a larger dispersion, which is beneficial to measure the energy spread. A quadrupole magnet, a fluorescent screen, and a SBPM with eight electrodes is installed near the cavity for calibration and comparison. The systematic framework and simulation results are also discussed in this paper.

1. Introduction

Energy spread is an important parameter that reflects the quality of the beam. Especially in linear accelerators used as storage rings and collider injectors, ensuring a good energy spread of the beam can improve injection efficiency and maintain beam stability. Therefore, real-time and high-resolution measurement of energy spread needs to be achieved. The most widely used method for energy spread measurement at present is to insert a fluorescent screen or an OTR after bending magnets. By measuring the generated light spot, the transverse distribution of the beam can be determined, and the energy spread is capable to be obtained based on the principle that different energy electrons have different turning radius [1]. However, this intercepting method is unsuitable for real-time measurement or feedback control. Therefore, a nonintercepting energy spread monitor is required. A feasible nondestructive method is using stripline beam position monitor (SBPM) [2, 3, 4]. SBPM is installed at positions with high value of dispersion function, allowing extraction of the beam's quadrupole moment to obtain the energy spread. The sensitivity of this method is influenced by the angle of electrodes.

In order to achieve the energy spread measurement with high-resolution, this paper proposes the method of RF resonant cavity. Similar to SBPMs, cavities can also measure the quadrupole moment, thus obtaining the energy spread. Compared to SBPMs, cavities have higher signal-to-noise ratio (SNR), enabling better quadrupole moment resolution, which can also be confirmed in the position measurement. Additionally, in the direction of the beam, cavities can realize a more compact layout.

This paper introduces a design of cavity-based quadrupole moment and energy spread monitor for the Dalian Coherent Light Source (DCLS), an extreme ultra-violet FEL facility with a length of 100 m, and provides simulation results to verify the feasibility of the proposed approach.

2. Theoretical basis

While passing through the cavity, the bunch will excite different eigenmodes. The signal amplitude of an eigenmode can be written as

$$V = k_n q \int_{-T}^{T} f(t) \exp(i\omega t)\, dt \ , \tag{1}$$

where q is bunch charge, the final integral is related to the longitudinal distribution of the bunch, and k_n is the loss factor, which is related to the mode type. As shown in Fig. 1, for a rectangular cavity rotated 45° about the axis, its TM220 mode amplitude is proportional to the beam quadrupole moment [5, 6].

$$V_{TM220} \propto x^2 - y^2 + \sigma_x^2 - \sigma_y^2 \ , \tag{2}$$

where x and y are the beam positions in the two directions. σ_x and σ_y are the beam rms sizes in the two directions. When the bunch length is short and does not change significantly, the quadrupole moment $x^2 - y^2 + \sigma_x^2 - \sigma_y^2$ can be obtained by extracting the charge and the amplitude of the TM220 mode. Assuming that the vertical dispersion, η_y, is zero, the relationship between energy spread and quadrupole moment can be written as

$$\sigma_x^2 - \sigma_y^2 = \beta_x \epsilon_x + (\eta_x \frac{\Delta E}{E})^2 - \beta_y \epsilon_y + g \ , \tag{3}$$

where β_x and β_y are β functions, η_x is the horizontal dispersion, ϵ_x is the horizontal emittance, and g is the correction factor, which can be determined by calibration. The two directions x and y are interchangeable. Therefore, after excluding the influence of the beam position, the relationship between the TM220 amplitude and the energy spread can be established. As indicated in Eq. (3), placing the monitor at a position with high value of dispersion function is beneficial to improve the sensitivity and resolution of the quadrupole moment measurement. For a linear accelerator, it is more suitable to place the monitor after the bending magnets.

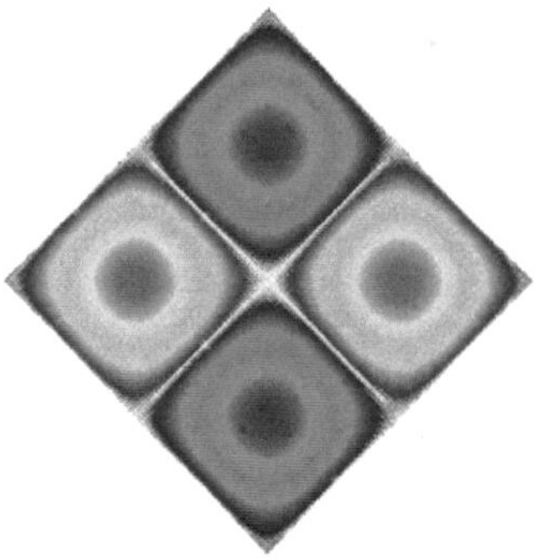

Figure 1. Electric field of TM220 mode in rectangular cavity. Warmer colors represent stronger electric fields.

Table 1. Beam Parameters in the Undulator Section of DCLS

Parameter	Value
Beam energy	300 MeV
Charge	500 pC
Bunch length (rms)	˜0.12 mm
Transverse dimension (rms)	˜0.1 mm
β function (mean)	6-29 m
Normalized emittance	1-2 mm·mrad

3. Layout and framework

The scheme for measuring beam energy spread using a cavity will be carried out in the undulator section of DCLS. The beam parameters in this section are shown in Table 1.

The cavity will be installed after the bending magnet at the end of the undulator section, as shown in Fig. 2. The dispersion at this position is significant, which is advantageous for energy spread measurement. A fluorescent screen is set downstream to get information of beam position and provide online calibration, and the upstream ICT on the bending magnet can synchronously provide absolute measurement of the beam charge. Additionally, an eight-electrode SBPM will be installed for energy spread measurement as a reference, which can also provide information of beam charge and beam position.

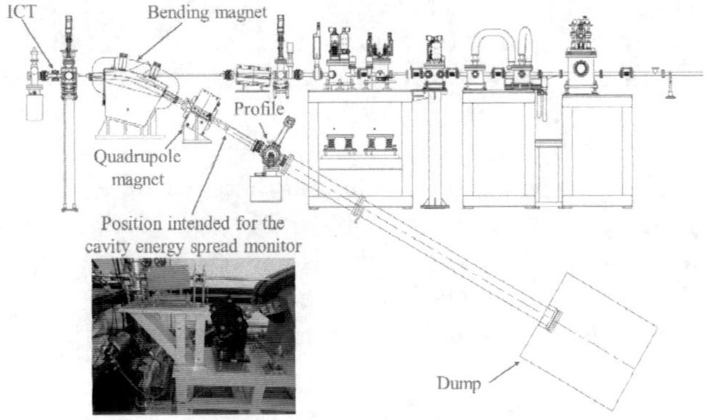

Figure 2. Layout diagram.

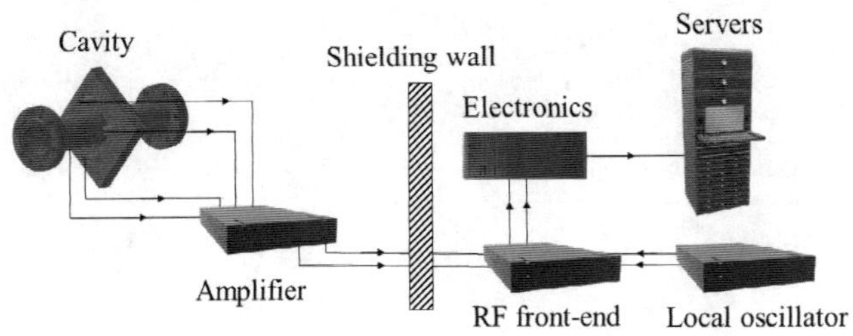

Figure 3. Systematic framework of the cavity quadrupole moment and energy spread monitor.

The systematic framework is shown in Fig. 3. When the beam passes through the cavity, the signal generated by the excited eigenmodes will be coupled out via feedthroughs. In order to overcome cable attenuation, the signal first enters an amplifier, then passes through long cables to the accelerator tunnel and connects to the analog front end for down conversion. Afterwards, it is fed into the electronics for amplitude extraction and energy spread calculation.

4. Physical design

The TM220 mode of the rectangular cavity can indicate the magnitude of the quadrupole moment. In order to share the analog front end with other cavity monitors on the DCLS, the operating frequency of the TM220 mode is set to be 4.76 GHz.

The relationship between the frequency of the TMmn0 mode and the dimensions of the cavity can be expressed as

$$f_{TM220} = \frac{c}{2}\sqrt{(\frac{m}{a})^2 + (\frac{n}{a})^2}\,, \tag{4}$$

where c is speed of light in vacuum, and a represents the length of one side of the square cavity. By calculation, we have $a = 89.1$ mm. In order to reduce the impact on beam, the longitudinal length l of the cavity is set to 10 mm.

The TM220 mode signal is coupled out by feedthroughs. The feedthroughs are set at the position with the strongest electric field of the TM220 mode, and are inserted along the electric field direction for optimal coupling. Four feedthroughs are used to maintain electric field symmetry. The inner conductors of the feedthroughs should not be inserted too deeply to avoid distorting the electric field.

The diameter of the beam pipe at the installation location is 35 mm. The cutoff frequency of it is 5.023 GHz, which ensures that the TM220 mode signal operating at 4.7 GHz does not leak out. The beam pipe will affect the frequencies of all eigenmodes inside the rectangular cavity.

When machining a rectangular cavity, the milling cutter cannot process the edge of the beam direction to 90°, so chamfering must be considered. A square groove is milled on a cubic metal plate to serve as the cavity. Due to the length of the cavity in the z-direction (beam direction) reaching 10 mm, a milling cutter with a blade length of 15 mm and a blade diameter of 6 mm is chosen to enter along the z-direction for cutting the groove. This will leave chamfers with a radius of 3 mm at the four corners. This also affects the frequencies of the eigenmodes. Therefore, numerical calculations are needed to correct the cavity dimensions obtained from Eq. (4). CST Microwave Studio [7] is used for frequency simulation calculation. The final structure and dimensions of the cavity are determined as shown in Fig. 4 and Table 2.

Table 2. Optimized Dimensions of the Cavity

Parameter	Value
a	86.2 mm
l	10 mm
p_y	61.8 mm
d	1.4 mm
Chamfer	R3

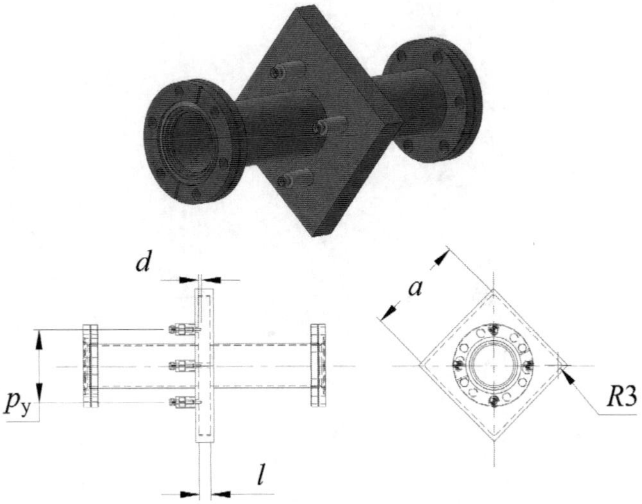

Figure 4. The structure of the rectangular cavity.

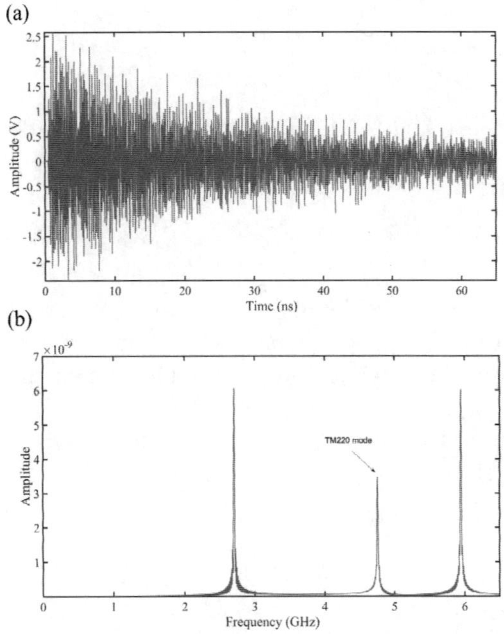

Figure 5. The output signal of the cavity when $x = 10, y = 0, \sigma_x = 0$, and $\sigma_y = 0$. (a) is the time-domain signal. (b) is the frequency-domain signal.

5. Simulation

The cavity model is established in CST, and virtual beams are loaded for simulation. The beam simulation is divided into two cases. In the first case, the transverse sizes of the beam are zero ($\sigma_x = 0, \sigma_y = 0$), and the beam position is changed. In the second case, the beam is located at the axis ($x = 0, y = 0$), and the transverse sizes are varied. The time-domain and frequency-domain output signals of the cavity are shown in Fig. 5.

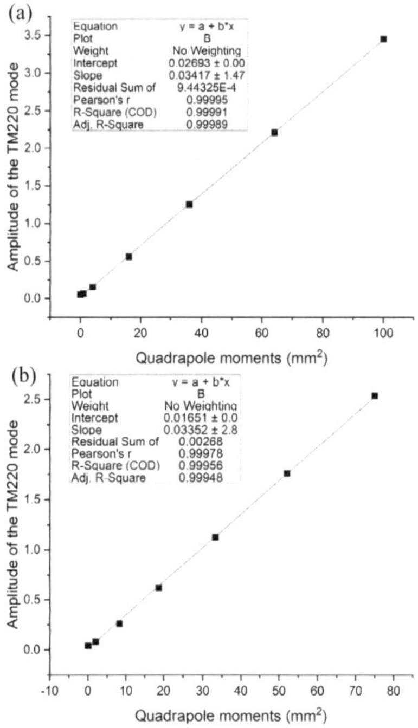

Figure 6. Variation of TM220 mode amplitude with beam quadrupole moment. (a) is x varying from 1 to 10 mm when $\sigma_x = 0, \sigma_y = 0$, and $y = 0$. (b) is σ_y varying from 0.03 to 8.66 mm when $x = 0, y = 0$, and $\sigma_x = 1$ mm.

The results show that there is a good linear relationship between the amplitude of the TM220 mode and the quadrupole moments. The quadrupole moment sensitivity of the cavity monitor is better than 0.03 mm^2 in the frequency domain. This parameter of the eight-electrode SBPM is generally only on the order of 10^{-3}. The cavity monitor is more likely to achieve better measurement resolution.

The simulation results show that the TM220 mode is successfully excited when the quadrupole moment is present. The variations of quadrupole moments $x^2 - y^2 + \sigma_x^2 - \sigma_y^2$ and the amplitude of TM220 modes obtained through simulation at different beam positions and beam transverse sizes are shown in Fig. 6.

6. Summary

In this paper, the design of the cavity quadrupole moment and energy spread monitor is shown, and the simulation results indicate a good linear relationship between its TM220 mode output signal and the beam quadrupole moment. The beam experiment will be carried out at DCLS soon. This study provides a new approach for nondestructive and high-resolution measurement of beam energy spread.

Acknowledgments

Acknowledgments This work is supported by the National Key Research and Development Proram of China No.2022YFA1602201, the International Partnership Program of the Chinese Academy of Sciences Grant No. 211134KYSB20200057 and the STCF Key Technology Research Project.

References

[1] Satoh M, Furukawa K and Suwada T 2005 *Proc. 2005 Particle Accelerator Conf. (PAC'05)* (Knoxville, TN, USA) pp 4212–4214
[2] Suwada T, Satoh M and Furukawa K 2003 *Phys. Rev. ST Accel. Beams* **6** 032801
[3] Dooling J C, Brumwell F R, Donley L, McMichael G E and Stipp V F 2004 *Proc. LINAC'04* (Lübeck, Germany) pp 405–407
[4] Fang J, Lu P, Luo Q, Sun B and Wang X H 2010 *Proc. IPAC'10* (Kyoto, Japan) pp 1029–1031
[5] Whitum D and Kolomensky Y 1999 *Rev. Sci. Instrum.* **70** 2300
[6] Luo Q, Zhou Z R, Li H T, Sun B G, Zheng J, Tang K and Lu P 2014 *Proc. IPAC'14* (Dresden, Germany) pp 3578–3580
[7] CST , https://www.3ds.com/

Microbunching of thermionic cathode rf gun beams in the Advanced Photon Source s-band linac

J Dooling, A Brill, I Lobach, A Lumpkin, N Kuklev, N Sereno, Y Sun
Argonne National Laboratory, 9700 S. Cass Ave., Lemont, IL 60439, USA
E-mail: dooling@anl.gov

Abstract. We report on measurements of beams from thermionic cathode (TC) rf guns in the Advanced Photon Source S-Band Linac. These measurements include the macropulse out of both new and existing TC guns as well as the observation of microbunching within the micropulses of these beams. A gun chopper limits the macropulse FWHM duration to the 10-ns range. Our objectives were to analyse the new TC gun and investigate microbunching within a TC-rf-gun-generated beam. Our diagnostics elucidated longitudinal beam structures from the ns to the fs time scales. Coherent transition radiation (CTR) interferometers responding to far-infrared wavelengths were employed after each compression stage to provide the autocorrelations of the sub-ps micropulse durations. The first compression stage is an alpha magnet and the second a chicane. A CCD camera was used to image the beam via optical transition radiation from an Al screen at the end of the linac and also employed to measure coherent optical transition radiation (COTR) in the visible range. The COTR diagnostic observations, implying microbunching on a fs time scale, are presented and compared with a longitudinal space-charge impedance model.

1. Introduction

The injection system of the Advanced Photon Source (APS) has relied on thermionic cathode (TC) rf guns as electron-beam sources since 2001 [1]. This will continue to be the case for the APS Upgrade with the multi-bend achromatic magnet lattice installation in the 6-GeV storage ring currently in progress [2]. There are two such S-Band TC rf guns installed in the linac so that one is a “hot” spare. We report the basic testing of the beams from new generation TC rf guns manufactured by RadiaBeam Technologies (RBT) [3] in terms of the ≈10-ns long macropulse composed of micropulses with an S-Band repetition frequency (2856 MHz). We obtained basic macropulse data for a newly installed RBT gun to comparison with one of a set of three TC rf guns purchased from AET Associates (AET) in 2001. The AET guns have been the primary electron source for the APS since that time.

Our interest was to measure the charge of each micropulse within the macropulse structure from the chopper-gated [4], TC-rf gun [5–7]. The extraction of charge from the TC rf gun involves high-power rf applied to the cathode; thus, the extracted charge comes in a series of micropulses. With the macropulse temporal distribution, we can then determine the charge per micropulse. Only the AET gun was employed for the bunch compression experiment discussed below.

Journal of Physics: Conference Series **2681** (2024) 012009 doi:10.1088/1742-6596/2681/1/012009

In the APS linac, the micropulse duration after the alpha magnet is sub-ps. After further compression in the chicane, longitudinal space charge (LSC) induced microbunching within the micropulses generates coherent optical transition radiation (COTR) when striking the downstream intercepting Al screen. The COTR (visible wavelengths, fs temporal scale) was transported out of the tunnel to a CCD camera, and the images were recorded. We present a comparison of the observed COTR enhancements with the predictions of a LSC-impedance model.

2. Experimental description

A schematic of the linac with components relevant for the present discussion is presented in Figure 1. Compression is provided by the alpha magnet at the end of the gun beam line and

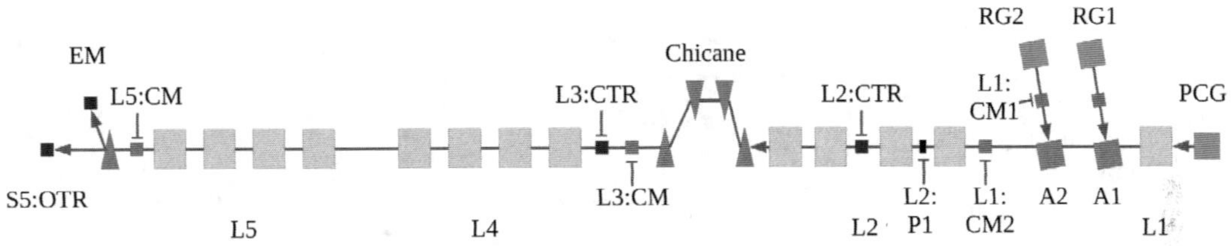

Figure 1. Linac schematic. The main accelerating sections are designated LX where X=1,2,4, and 5. Sections 2, 4, and 5 each have four 3-m SLEDed S-band accelerating structures. A1 and A2 are the alpha magnets associated with two TC rf gun beamlines RG1 and RG2. L2:P1 represents the BPM location for the macropulse measurements. L*:CM are current/charge monitors. EM is a spectrometer, and triangles represent dipole magnets. OTR images are collected at the S5:OTR station.

the L3 chicane after the first four accelerating structures in L2. Coherent Transition Radiation far-infrared (CTR FIR) interferometers, positioned in the L2 section (L2:CTR) and after the chicance (L3:CTR), provide longitudinal pulse length data [8]. Average current is monitored immediately after the gun and macropulse charge is measured after the alpha magnet in the linac beamline. Optical transition radiation (OTR) and COTR are generated by an Al screen (S5:OTR, as shown in Figure 1) when struck by the electron beam. An optical transport line guides the light to a camera system outside of the tunnel.

A block diagram of the macropulse duration measurement is presented in Figure 2. Raw beam signals from the four button pickups are input to 20-dB directional couplers (Mini-Circuits ZGDC20-33HP+). The coupled signals ("c") are combined to form the sum output (Mini-Circuits ZN4PD-642W-S+) which is fed into one channel of a fast oscilloscope with $50-\Omega$ input impedance. The main, uncoupled signals continue on to the analog conditioning electronics associated with the rest of the S-band linac BPMs [9–11]. Because of capacitive coupling, the DC component of the macropulse is not present. Alternatively, the summing unit can be bypassed to provide coupled input to the oscilloscope from all four of the BPM channels. Four channel input allows for observation of beam position within the macropulse. The BPM rf signals are digitally undersampled by the oscilloscope's analog-to-digital converter following the discussion in [12].

A new 1.5-cell, TC rf gun developed by RBT was installed at the RG1 location in the linac during the December 2022-January 2023 APS maintenance period. We wished to measure the macropulse waveform of this gun to compare with that of the operating TC gun. Further tests with the RBT gun will be conducted once accelerator operations resume in Fall 2023. Only the AET gun, installed in the RG2 location, was used in the bunch compression experiment discussed below.

Journal of Physics: Conference Series **2681** (2024) 012009 doi:10.1088/1742-6596/2681/1/012009

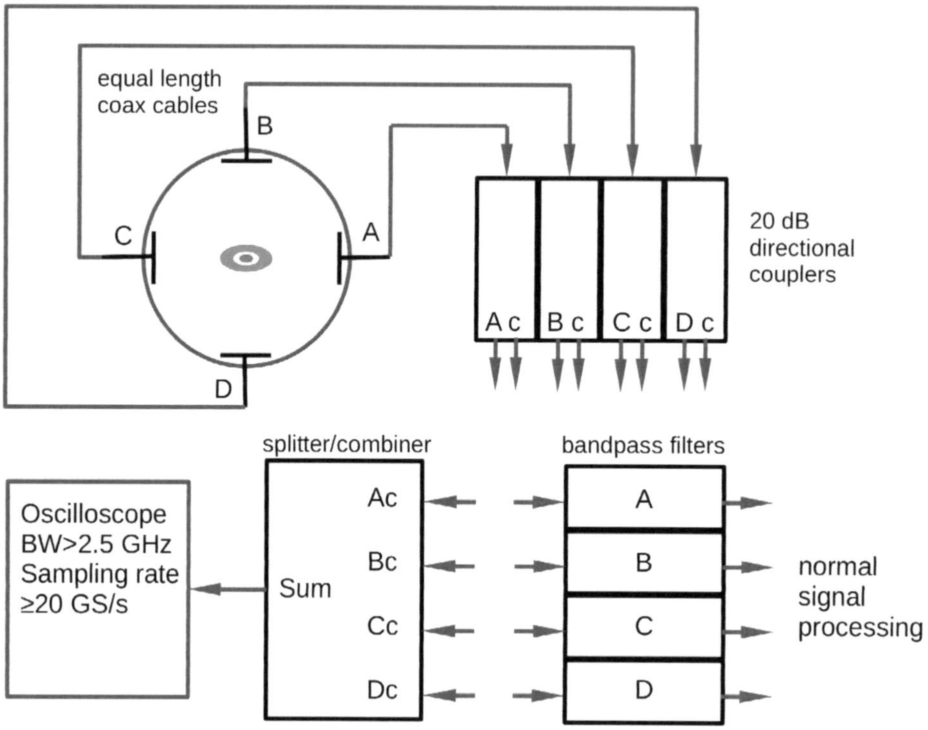

Figure 2. Macropulse measurement schematic.

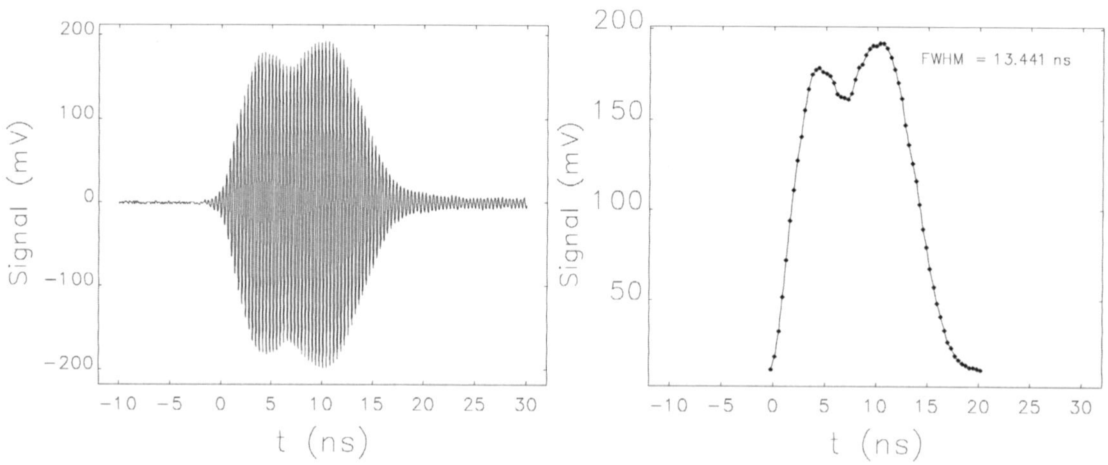

Figure 3. Left: Raw macropulse signal from the AET RG2 gun. Right: Peak detect envelope of the same signal.

3. Macropulse waveforms

An example of a raw macropulse waveform from one of the AET guns is plotted in Figure 3, left. Figure 3, right shows the same signal after peak-detection processing to yield the envelope. This AET gun is located in the RG2 station; this is the gun beamline closest to the entrance of the linac known as the L2 section. A sampling rate, f_s of 20 GS/s provides 7 samples per fundamental rf period (350 ps) and yields a relatively smooth envelope after peak detection.

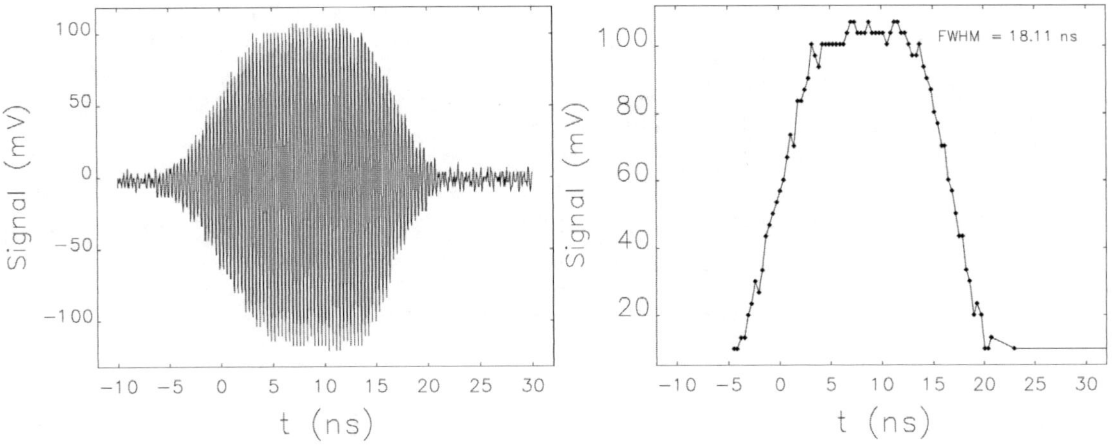

Figure 4. Left: Raw macropulse signal from the RG1 RadiaBeam gun. Right: Peak detect envelope of the same signal.

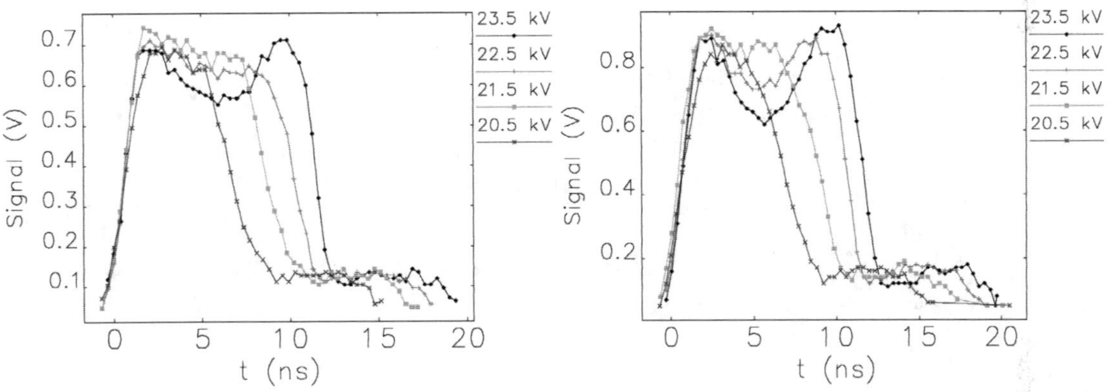

Figure 5. Macropulse envelope waveforms from the RG2 AET gun with varying chopper voltages for 190 mA (left) and 300 mA (right).

Macropulse data acquired for one of the new RadiaBeam guns is presented in Figure 4. In this case, the gun is located in the RG1 station, approximately 3 m further from the L2 linac section than the RG2 gun. Transmission losses RG1 are typically higher than for RG2. Each station has its own chopper system; in the case of RG1 the chopper pulse is longer than that for RG2.

The macropulse waveforms were also observed as the chopper voltage as well as the average gun output current were varied. Waveforms at four chopper voltages are presented in Figure 5 left and right for gun currents of 190 and 300 mA, respectively. Cathode heater power was varied to change the average gun current for the waveforms presented in Figure 5. At higher voltages, we observed a double peak in the macropulse envelope. In this case, the beam is swept past the slit on the rising edge of the voltage pulse and again on the falling edge. Gun chopper voltage is the method used to control the charge per macropulse in the linac while the gun cathode heater power is typically maintained at a constant value. Macropulse duration varies from approximately 7 to 11 ns for both current levels as the chopper is raised from 20.5 to 23.5 kV. Current and charge data collected from current monitors in the linac during these measurements are presented in Table 1.

Table 1. RG2 current and charge measurements

Chopper voltage (kV)	190 mA		300 mA	
	I_{CM1} (mA)	Q_{CM2} (nC)	I_{CM1} (mA)	Q_{CM2} (nC)
20.5	191	0.85	303	1.04
21.5	189	1.07	302	1.37
22.5	191	1.26	303	1.56
23.5	191	1.37	300	1.64

4. Micropulses and microbunching

The gun output current after the alpha magnet is modeled as a train of charge delta functions where each charge pulse is separated from its neighbor by one rf period. The current waveform may be expressed as,

$$I_g(t) = A(t)\delta\left(t - t_s - \frac{2\pi n}{\omega_{rf}}\right), \tag{1}$$

where $A(t)$ is the time-varying amplitude of the macropulse, t_s is time slip or jitter, δ is the delta function, and ω_{rf} is the radian frequency of RG2 rf drive power. The total charge is then determined integrating Equation 1 over time,

$$Q = \int I_g(t)dt = \sum_n A(t_n), \tag{2}$$

where $t_n = t_s + 2\pi n/\omega_{rf}$. Respective calibration factors for the AET RG2 gun operating at currents of 190 mA and 300 mA are 56.33 $\pm$0.97 pC/V and 55.36 $\pm$1.49 pC/V; the maximum charge per micropulse at these currents are 39.99 $\pm$0.88 pC and 49.88 $\pm$1.66 pC.

For bunch compression studies, we are interested in the micropulse peak current. Compression studies were conducted with the RG2 AET TC gun beam. Two coherent transition radiation (CTR) interferometers are installed in the APS linac [8, 13]; one in the L2 section (L2CTR) between the second and third accelerating structures (AS), and another following the chicane in L3 (L3CTR), see Figure 1. The four AS of L2 may be phased to chirp the pulses prior to the chicane to facilitate bunch compression. The beam energy after the L2 section was 150 MeV. The L2CTR sees the effect of the alpha magnet compression, whereas L3CTR observes compression provided by the chicane. Autocorrelations of the CTR signals at these two locations are presented in Figure 6. Assuming a Gaussian profile for the compressed bunch, the rms bunch duration can be approximated as $\sqrt{2}\sigma_z/c$. For the autocorrelation scans shown in Figure 6 this yields 300 fs and 130 fs from L2CTR and L3CTR, respectively. We assumed that the micropulse length is approximately constant over the macropulse duration. Using the FWHM duration, 50 pC results in a current of 170 A after the L3CTR. This current level is consistent with predictions made in [5].

During bunch-compression studies, gun output current was increased to 650 mA. Phasing the L2 AS to maximize compression, COTR was observed on the S5:OTR monitor, at an energy of 375 MeV [14]. Examples of OTR and COTR emission are presented in Figure 7. Emission intensity contour and profile data with L2 rf phased away from maximum compression are shown in the top row; the bottom row presents enhanced emission (COTR) at the maximum compression phase.

The macropulse beam charge was monitored at L5:CM during data acquisition and was stable to less than 3 percent for both compressed and uncompressed measurements. The

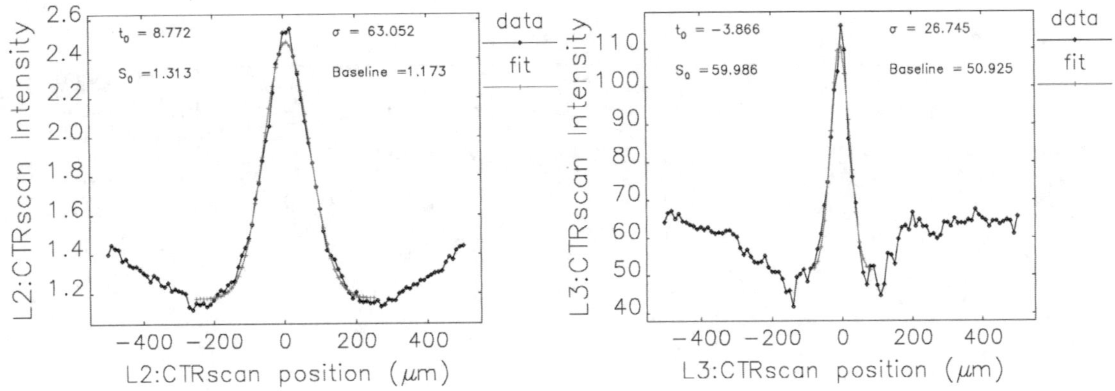

Figure 6. Autocorrelations of the coherent transition radiation signals from the L2CTR (left) and L3CTR (right) along with Gaussian fits to the central peak. RG2 (AET) source.

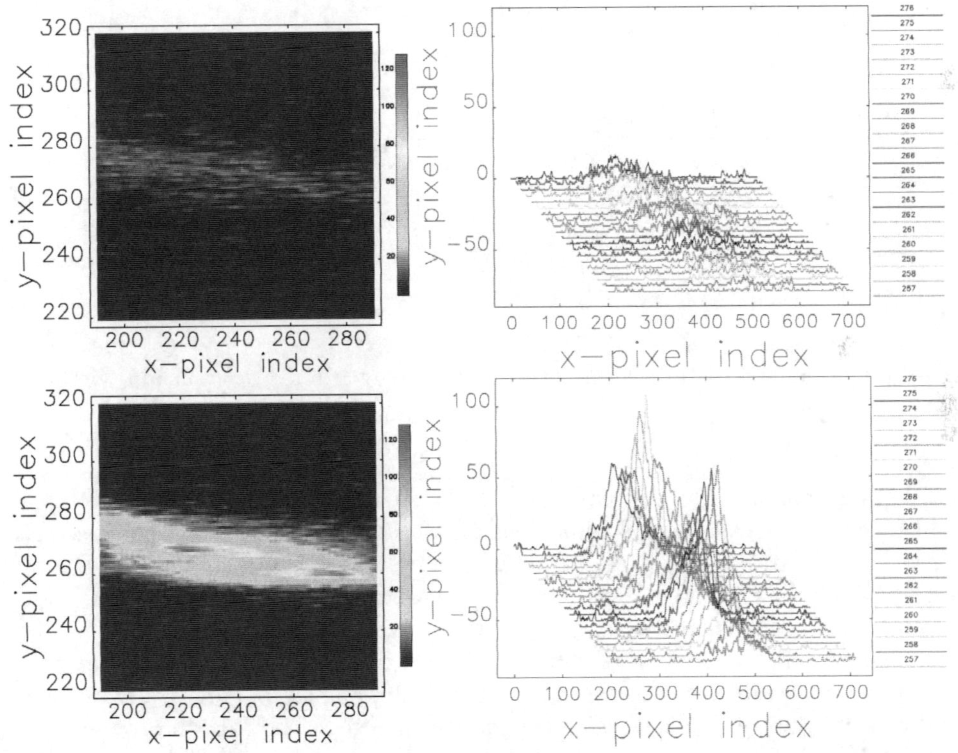

Figure 7. Top: Intensity contour plot (left) and line profiles (right) away from peak compression. Bottom: Intensity contour plot and line profiles near peak compression.

COTR/OTR intensity ratio is about 6.5 at the most enhanced spatial location. The two COTR-enhanced locations in the lower image (maximum compression) are approximately σ_x=380 μm by σ_y=50 μm in size. Compressed micropulse peak current was approximately 200 A.

Examining the effect of longitudinal space-charge (LSC) impedance, we plot in Figure 8 Equation 26 from [15] for R_{56}=65 mm. With an rms energy spread of 2 keV, a gain of 5.7 is obtained at 800 nm. Signal intensity is proportional to the product of the COTR gain and the sensitivity of the CCD sensor integrated over the wavelength range of interest (400-1000 nm).

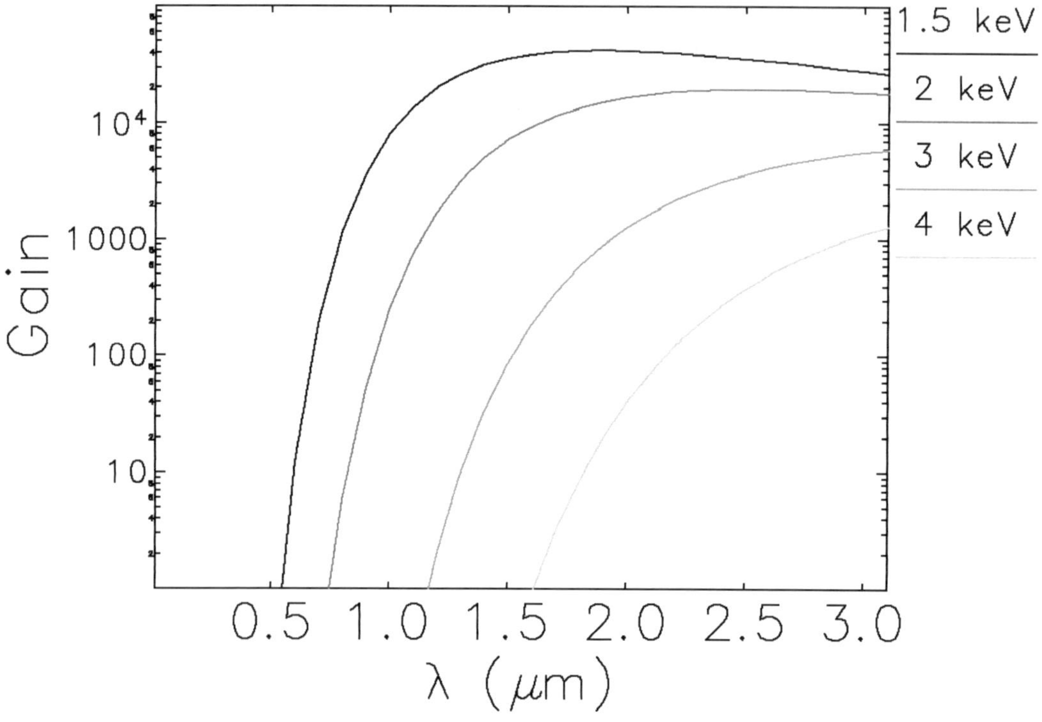

Figure 8. Microbunching gain as a function of wavelength for $R_{56} = 65\,\mathrm{mm}$ with rms energy spread as a parameter.

Picking a COTR gain value in the red end of the range is meant to represent a rough estimate of the enhancement ratio. The gain is in reasonable agreement with the enhancement factor determined for the data plotted in Figure 7. This is the first time a comparison has been done of empirical COTR gain from a TC rf gun source with an LSC impedance model.

Previous simulations of the LSC instability showed that microbunching in the visible wavelength regime can be supported if the slice energy spread of the microbunched portion of the beam is approximately 2-4 keV [16]. Using a PC rf gun source, an LCLS-I experiment generated COTR gain consistent with a slice energy spread of 3 keV. In the context of Ratner's LSC model, a 2-keV slice energy spread would be indicated to explain the observed COTR gain for the microbunched portion of the TC rf gun beam. This value is much smaller than the measured ensemble energy spread.

Broadband visible regime COTR enhancements have also recently been reported in laser-plasma-wakefield-accelerated beams [17–19], however this is a different microbunching mechanism. In this case, intense fields from a 100-TW-class drive laser interacting with the plasma electrons lead to microbunching.

5. Summary

We described a method to measure the macropulse waveform of the APS TC rf guns. This method requires an oscilloscope with bandwidth and sampling rate sufficient to monitor the fundamental frequency of the rf applied to extract beam from the gun; in the present case, this requires a bandwidth > 2.5 GHz and sampling rate > 20 GS/s for 2865 MHz S-band rf. Measurement of the macropulse envelope allows for calibration of the microbunch charge and thereby calculation of peak currents after bunch compression. AET and RadiaBeam TC guns were tested and, accounting for differences in the chopper systems, behaved in a similar manner. Other methods employing diodes could be explored to eliminate the cost of an oscilloscope [20].

Additionally, we have inferred the presence of LSC-induced microbunching at visible wavelengths via the COTR mechanism from TC rf gun beams. After two stages of compression and with further acceleration to 375 MeV, an enhancement factor of 6.5 was observed in OTR near maximum compression. The enhanced emission is attributed to the generation of COTR from microbunching in the TC rf gun-generated micropulses. Analytical models indicate that a slice energy spread of only a few keV is needed for microbunching (COTR gain) at visible wavelengths where the CCD camera is sensitive. The importance of slice energy spread as it applies to a TC-rf gun-generated beam is analogous to what was reported for PC gun [16] and LPWA [17] beams. Further investigations including spectral measurements of the microbunching phenomena are warranted.

Acknowledgements
Thanks to R. Soliday and H. Shang of APS ANL for assistance with analysis scripts and to K. P. Wootton for editorial help.

References

[1] Beczek K, Lewellen J, Nassiri A, and Tanabe E, A Rationalized Approach to Thermionic RF Gun Design, Proc. PAC'01, WPAH052, Chicago, IL, USA, June 2001, JACoW Publishing, Geneva, Switzerland, pp 2206–2208, https://doi.org/10.1109/PAC.2001.987325

[2] Advanced Photon Source Upgrade Project Final Design Report, APSU-2.01-RPT-003, Advanced Photon Source, May 2019, p 2, https://publications.anl.gov/anlpubs/2019/07/153666.pdf

[3] Kutsaev S *et al*, A New Thermionic RF Electron Gun for Synchrotron Light Sources, Proc. NAPAC'16,TUB3CO04, Chicago, IL, USA, Oct. 2016, JACoW Publishing, Geneva, Switzerland, pp 453–456, https://doi.org/doi.org/10.18429/JACoW-NAPAC2016-TUB3CO04

[4] Kang Y, Wang J, Milton S, Teng L, Thompson K, Jones J, Walters D, and Jagger J, Beam Chopper for the Low-Energy Undulator Test Line (LEUTL) in the APS, Proc. PAC'97, pp 2699–2701, 6W001, Vancouver, Canada, May 1997, JACoW Publishing, Geneva, Switzerland https://accelconf.web.cern.ch/pac97/papers/pdf/6W001.PDF

[5] Borland M, 1991, A High-Brightness Thermionic Microwave Electron Gun, Stanford University, Available as SLAC Report 402, https://www.slac.stanford.edu/pubs/slacreports/reports03/slac-r-402a.pdf

[6] Borland M, An Improved Thermionic Microwave Gun and Emittance-Preserving Transport Line, Proc. PAC'93, pp 3015–3017, Washington D.C., USA, Mar. 1993, JACoW Publishing, Geneva, Switzerland, https://accelconf.web.cern.ch/p93/PDF/PAC1993_3015.PDF

[7] Lewellen J, Biedron S, Lumpkin A, Milton S, Nassiri A, Pasky S, Travish G, and White M, Operation of the APS RF Gun, Proc. LINAC'98, pp 863–865, TH4042, Chicago, IL, USA, Aug. 1998, JACoW Publishing, Geneva, Switzerland, https://jacow.org/l98/papers/TH4042.pdf

[8] Lumpkin A, Berg W, Borland M, Lewellen J, and Sereno N, Initial CTR-Based Bunch Length Measurements of Linac Beams Following the APS Bunch Compressor, Proc. FEL'03, pp 2420–2422, WPPB005, Tsukuba, Japan Sept. 2003, JACoW Publishing, Geneva, Switzerland, https://accelconf.web.cern.ch/p03/PAPERS/WPPB005.pdf

[9] Grelick A, RF and Beam Diagnostic Instrumentation at the Advanced Proton Source (APS) Linear Accelerator, Proc. LINAC'96, pp 851–853, THP75, Geneva, Switzerland, Aug. 1996, JACoW Publishing, Geneva, Switzerland, https://jacow.org/l96/papers/THP75.pdf

[10] Lill R, Singh O, and Arnold N, New Beam Position Monitor System Design for the APS Injector, AIP Conference Proceedings, **648**, 1, pp 401–408, 2002, 0094-243X, https://doi.org/10.1063/1.1524426

[11] Sereno N, Borland M, and Lill R, 2008, Automated correction of phase errors in the Advanced Photon Source linac, *Phys. Rev. ST Accel. Beams*, **11**, 7, 072801, American Physical Society, https://doi.org/10.1103/PhysRevSTAB.11.072801

[12] Wendt M, 2014, Direct (Under)Sampling vs Analog Downconversion for BPM Electronics, Proc. IBIC'14,

pp 486–494, WEIYB1, Monterey, CA, USA, Sep. 2014, JACoW Publishing, Geneva, Switzerland, https://accelconf.web.cern.ch/IBIC2014/papers/WEIYB1.pdf

[13] Lumpkin A, Yang B, Berg W, Lewellen J, Sereno N, and Happek U, 2000, Electron beam bunch length characterizations using incoherent and coherent transition radiation on the APS SASE FEL project, *Nuclear Instruments and Methods in Physics Research A*, **445**, pp 356–361, Elsevier Science.

[14] Lumpkin A, Beam Diagnostics for Coherent Optical Radiation Induced by the Microbunching Instability, Proc. FEL'13, pp 169–172, TUOBNO01, New York, NY, USA, Aug. 2013, JACoW Publishing, Geneva, Switzerland, https://accelconf.web.cern.ch/FEL2013/papers/TUOBNO01.pdf

[15] Saldin E, Schneidmiller E, and Yurkov M, 2002, Klystron instability of a relativistic electron beam in a bunch compressor, *Nuclear Instruments and Methods in Physics Research A*, **490**, pp 1–8, Elsevier Science, https://doi.org/10.1016/S0168-9002(02)00905-1

[16] Ratner D, Chao A, and Huang Z, Three-dimensional Analysis of Longitudinal Space Charge Microbunching Starting from Shot Noise, Proc. FEL'08, pp 338–341, TUPPH041, Gyeongju, Korea, Aug. 2008, JACoW Publishing, Geneva, Switzerland, https://accelconf.web.cern.ch/FEL2008/papers/TUPPH041.pdf?n=FEL2008/papers/TUPPH041.pdf

[17] Lin C *et al*, 2012, Long-Range Persistence of Femtosecond Modulations on Laser-Plasma-Accelerated Electron Beams, *Physical Review Letters*, **108**, 094801, American Physical Society, https://doi.org/10.1103/PhysRevLett.108.094801

[18] Lumpkin A *et al*, 2020, Coherent Optical Signatures of Electron Microbunching in Laser-Driven Plasma Accelerators, *Physical Review Letters*, **125**, 014801, APS, https://doi.org/10.1103/PhysRevLett.125.014801

[19] LaBerge M *et al*, Coherent 3D Microstructure of Laser-Wakefield-Accelerated Electron Bunches, Proc. FEL'22, THAI2, Trieste, Italy, Aug. 2022, JACoW Publishing, Geneva, Switzerland, https://indico.jacow.org/event/44/contributions/347/editing/paper/685/2263/THAI2.pdf

[20] Gasior M, First Results from the LHC Schottky Monitor Operated with Direct Diode Detection, Proc. BIW'12, pp 222–224, TUPG044, Newport News, VA, USA, Apr. 2012, JACoW Publishing, Geneva, Switzerland, https://accelconf.web.cern.ch/BIW2012/papers/tupg044.pdf

AUTHOR INDEX